OXFORD MEDICAL PUBLICATIONS

Textbook of Medical Record Linkage

TEXTBOOK OF MEDICAL RECORD LINKAGE

EDITED BY

J. A. BALDWIN (deceased), MA, MD, FFCM, FRC PSYCH.
Formerly Director of Oxford Record Linkage Study

E. D. ACHESON DM, FRCP, FFCM
Chief Medical Officer, Department of Health and Social Security
(Former Director of Oxford Record Linkage Study)

AND

W. J. GRAHAM D.PHIL.
Research Fellow, London School of Hygiene and Tropical Medicine
(Former Research Officer, Oxford Record Linkage Study)

OXFORD NEW YORK TOKYO
OXFORD UNIVERSITY PRESS
1987

Oxford University Press, Walton Street, Oxford OX2 6DP

Oxford New York Toronto
Delhi Bombay Calcutta Madras Karachi
Petaling Jaya Singapore Hong Kong Tokyo
Nairobi Dar es Salaam Cape Town
Melbourne Auckland
and associated companies in
Beirut Berlin Ibadan Nicosia

Oxford is a trade mark of Oxford University Press

Published in the United States
by Oxford University Press, New York

British Library Cataloguing in Publication Data
Textbook of medical record linkage.—
(Oxford medical publications)
1. Medical records—Data processing
I. Baldwin, J. A.   II. Acheson, E. D.
III. Graham, W. J.
651.5'9   RA976
ISBN 0-19-261319-7

Library of Congress Cataloging in Publication Data
Textbook of medical record linkage.
(Oxford medical publications)
Includes bibliographies and index.
1. Medical record linkage.   I. Baldwin, J. A.
(John A.)   II. Acheson, E. D. (Ernest Donald), 1926–
III. Graham, W. J.   IV. Series. [DNLM: 1. Medical
Records. WX 173 T355]
R864.T49   1986   614.4'028   86-5340
ISBN 0-19-261319-7

Set by Joshua Associates Limited, Oxford
Printed in Great Britain by
St Edmundsbury Press
Bury St Edmunds, Suffolk

FOREWORD
SIR RICHARD DOLL

The human brain has a fantastic capacity to retain and recall information about past events. This enabled the great physicians of the past to define clinical syndromes, gain a picture of the natural history of individual diseases, and recognize associations that led eventually to an understanding of causation. They were, however, limited to their own experience, which could embrace only a small number of patients and a small part of those patients' lives. Note-taking helped by enabling one physician to share the experience of others; but its ability to do so was restricted by the physical dispersal of the notes and the time taken to search through them. With the coming of computers, the position was qualitatively changed. Information from many different sources could be centralized, the time taken to examine it was reduced to fractions of a second, and the opportunities for associating the medical events of a patient's life with his past experiences were enormously increased.

Three members of the Nuffield Department of Clinical Medicine in Oxford, Dr. Donald Acheson, Dr. Sidney Truelove, and the late Professor Leslie Witts quickly appreciated what the new technical developments were making possible, and in 1961 they wrote to the *British Medical Journal* suggesting that analysis of medical records linked nationally 'would show important clinical associations between one disease and another'. It would they added,

provide excellent morbidity statistics and create a science of prognosis. The study of human genetics would be transformed, provided the patients' identifying numbers were also inserted on birth certificates. These would be a base from which field studies of the epidemiology of the important diseases of the day would spring. We regard this last point as a cardinal one, because the chronic non-infective diseases have become the great medical problems of today and of the future, and one of the great difficulties with these diseases is the possibility that an environmental cause may be remote in time and space from the clinically recognizable disease.

The merits of the idea were immediately recognized by epidemiologists and forward-looking health service administrators; but its practice has proved harder than was at first envisaged. Lack of a personal identity number used both for social security and health service purposes and known to every individual has complicated the identification of records relating to one individual that are made in different places; the early computers were too slow and too small to handle the mass of information economically; and the possibility that centrally stored information about individuals might be used to their detriment (as has happened occasionally with data collected for commercial purposes) led to a heightened concern for ensuring that the confidentiality of the medical record was preserved and to resistance to its central storage.

In the face of these difficulties, progress was slow and it has not yet been possible to develop a national system of record linkage of the type that was originally envisaged in any country. Now, however, the position has changed. The pilot studies described in this volume have shown that the problem of identification can be overcome and that results can be obtained that are capable of providing new clues to causation. Modern techniques have ensured that the confidentiality of records held centrally in a computer can be protected more rigorously than written records stored in hospital or general practitioners' notes, and it has gradually been appreciated that bona fide medical research workers need access to such records if the public is to be protected against the hazards of industrial pollution, the side-effects of drugs, and the spread of infection, which the experience of AIDS has shown can still be felt as a major threat. The use of personal medical information for such purposes has long been the practice of the medical profession, whose ethical code has ensured that information shared for research purposes was treated in strict confidence, and it has now been enshrined in the 1984 Data Protection Act and the National Health Service's code of practice.

There remains the final hurdle: the extention of the system on a large enough scale to embrace all the important medical events of an individual's life; for it is only then that its full potentialities can be realized. This may not be possible in a single step in a country as large as Britain and it may be necessary to try it first on a regional scale. The results that Drs Baldwin, Acheson, and Graham have brought together provide strong grounds for giving such an extension high priority in the competition for a share of the national funds for health.

MEMOIR TO JOHN BALDWIN

When John Baldwin died tragically in his sleep at the age of 55 this book was already in preparation. It is hoped that it will be regarded as an appropriate memorial to his life's work on the use of medical records in epidemiological research.

John Baldwin entered medicine as a graduate in social sciences and the unravelling of the relationship of ill health and in particular of mental illness to social factors remained the enduring interest of his career. After graduating with Honours at Aberdeen University in 1952 he trained in psychiatry and became a Clinical and Research Fellow at Harvard University and the Massachusetts General Hospital. He focused on epidemiological techniques in psychiatry, and in particular on the definition of syndromes of mental illness and the measurement of their incidence. On his return to Scotland he remained in full-time research and founded the Aberdeen Psychiatric Case Register. A number of important publications on psychiatric epidemiology followed. He took his MD with Honours in 1969 and was elected FFCM in 1972 and FRCPsych. in 1976.

In his work as Director of the Oxford Record Linkage Study between 1968 and 1982 John Baldwin brought to bear not only his experience in the use of routine data but his considerable expertise in the use of computers in the manipulation of large files and in statistics. His reorganization of the files led to a number of original contributions including a study bearing on one of his major interests—the relationship of schizophrenia to physical illness.

John had a wide circle of friends and his sudden and untimely death brought forth a large number of tributes from colleagues in many different countries.† He will be remembered as a shy, modest, and thoughtful person and as a scientist who struggled for perfection in all his work. His friends were reminded of the depth of his thought and the breadth of interests in his final achievement. This was the preparation in the final year of his life of a generally accepted draft code of practice to deal with the previously intractable problem of balancing the need for medical records to remain private with the need of bona fide research workers to use them. If as seems likely his approach to this problem proves successful he will be remembered with gratitude by epidemiologists as the person who enabled epidemiology to move forward without undue constraint on a fair and firm foundation.

London   E.D.A.
September 1985

† A full tribute to J.A.B. is published in the *Journal of Community Psychology* **13**, 90–8 (1985).

ACKNOWLEDGEMENTS

The completion of this book would not have been possible without the co-operation and assistance of many people. Particular thanks are due to those who have contributed chapters, not the least for their patience during the lengthy preparation of the book.

Many of the chapters have arisen directly or indirectly from the work of the Oxford Record Linkage Study and a considerable number of people have contributed to this study during the last twenty or so years. The ORLS has received support from a variety of sources over this period, but in particular from the Department of Health and Social Security, the Nuffield Organization, the Nuffield Provincial Hospitals Trust, and the Oxford Regional Health Authority. In preparing the book, a special acknowledgement is due to Dr. Michael Goldacre, Director of the ORLS, Mr. Leicester Gill, Chief Computer Scientist to the ORLS, and Dr. Rosemary Rue, General Manager of the Oxford Regional Health Authority.

Finally, particular thanks are given to Mrs. Judy Brown for her secretarial assistance throughout.

The permission of the following to reprint material in this volume is gratefully acknowledged:

Memoir to John Baldwin   *Journal of Community Psychology* (1985) **13**, 90–8. Clinical Psychology Publishing Co.

Chapter 1   *American Journal of Human Genetics* (1967) **19**, 335–59, The University of Chicago Press, Chicago.

Chapter 4   *Journal of Epidemiology and Community Health* (1982) **36**, 69–79, British Medical Association.

Chapter 5   *British Journal of Industrial Medicine* (1972) **29**, 134–41, British Medical Association.

Chapter 6   *Journal of the American Medical Association* (1982) **248**, 1989–95, Copyright 1982, American Medical Association.

Chapter 7   From Hemmings, G. (1980). *The biochemistry of schizophrenia and addiction*, MTP Press, Lancaster.

Chapter 8   *British Journal of Preventive and Social Medicine* (1965) **19**, 164–73, British Medical Association.

Chapter 9   *Psychological Medicine* (1981) **11**, 341–50, Cambridge University Press.

Chapter 10   *Journal of Epidemiology and Community Health* (1981) **35**, 25–31, British Medical Association.

Chapter 11   *Human Toxicology* (1983) **2**, 63–73, Macmillan.

Chapter 12   *International Journal of Epidemiology* (1979) **8**, 305–12.

Chapter 13   *British Journal of Preventive Medicine* (1974) **28**, 104–7, British Medical Association.

Chapter 14   *British Journal of Preventive and Social Medicine* (1966) **20**, 49–57, British Medical Association.
Chapter 15   *Developmental Medicine and Child Neurology* (1976) **18**, 643–56, Spastics International Medical Publications.
Chapter 16   *British Medical Journal* (1983) **286**, 115–17 (Extracts), British Medical Association.
Chapter 17   *British Medical Journal* (1978) **1**, 83–4, British Medical Association.

CONTENTS

Obstetrics and child health

Adverse effects of drugs

Congenital disease and studies of the family

PART III   IMPLICATIONS OF RECORD LINKAGE

CONTRIBUTORS

DR. E. D. ACHESON,
Chief Medical Officer,
Department of Health and
    Social Security,
Alexander Fleming House,
Elephant and Castle,
London SE1, U.K.

SIR RICHARD DOLL,
Honorary Consultant,
ICRF—Cancer Epidemiology and
    Clinical Trials Unit,
Radcliffe Infirmary,
Oxford OX2 6HE, UK.

DR. D. S. FEDSON,
Associate Professor,
Head, Division of Internal Medicine,
Box 494,
University of Virginia Medical Centre,
Charlottesville,
Virginia 22908
USA.

PROFESSOR C. DU V FLOREY,
Department of Community Medicine,
Ninewells Hospital and Medical School,
Dundee DD7 9SY, UK.

PROFESSOR A. J. FOX,
Social Statistics Research Unit,
City University,
Northampton Square,
London EC1, UK.

MR L. E. GILL,
Chief Computer Scientist,
Oxford Record Linkage Study,
Oxford Regional Health Authority,

Old Road,
Headington,
Oxford OX3 7LF, UK.

DR. W. J. GRAHAM,
Research Fellow,
London School of Hygiene and Tropical
    Medicine,
Keppel Street,
(Gower Street),
London WC1E 7HT, U.K.

DR. M. J. GOLDACRE,
Director,
Oxford Record Linkage Study,
Oxford Regional Health Authority,
Old Road,
Headington,
Oxford OX3 7LF, UK.

DR. J. GOLDING,
Wellcome Senior Lecturer,
Department of Child Health,
University of Bristol,
Royal Hospital for Sick Children,
77 St. Michael's Hill,
Bristol BS2 8BJ, UK.

DR. G. R. HOWE,
Associate Professor,
Epidemiology Unit,
National Cancer Institute of Canada,
Faculty of Medicine,
McMurrich Building,
University of Toronto,
Toronto,
Ontario M5S 1A8,
Canada.

PROFESSOR R. E. KENDELL,
University Department of Psychiatry,
Royal Edinburgh Hospital,
Morningside Park,
Edinburgh EH10 5HF, UK.

DR. H. B. NEWCOMBE,
P.O. Box 135,
Deep River,
Ontario KOJ 1PO
Canada.

DR. M. L. NEWHOUSE,
Senior Research Fellow,
TUC Institute for Occupational Health,
London School of Hygiene and Tropical
   Medicine,
Keppel Street,
(Gower Street),
London WC1E 7HT, UK.

DR. J. E. OLIVER
Consultant Psychiatrist,
Burderop Hospital,
Wroughton,
Swindon,
Wiltshire SN4 0QA, UK.

PROFESSOR D. C. G. SKEGG,
Department of Preventive and Social
   Medicine,
University of Otago,
Dunedin,
P.O. Box 913,
New Zealand.

PROFESSOR M. P. VESSEY,
Department of Community Medicine and
   General Practice,
Radcliffe Infirmary,
Oxford OX2 6HE, UK.

Introduction

E. D. ACHESON

More than two decades have elapsed since the Oxford Record Linkage Study was set up in 1962. I was therefore grateful when John Baldwin gave me an opportunity to reflect briefly on developments in the field since that time, particularly the extent to which the potential uses of linked medical records have been realized.

The basic idea underlying record linkage is that two or more items of information about a person or a family recorded in different places may, when considered together, be of greater significance than when considered separately (Acheson 1967). I first became aware of this during a period of study at the US Veterans Administration Central Office in Washington, where abstracts of the hospital and death records of a large sample of US ex-servicemen were assembled. Examination of brief summaries of the successive hospital admissions (for quite another purpose) revealed the hitherto unknown link between regional enteritis (also ulcerative colitis) and ankylosing spondylitis (Acheson 1960). The conventional idea of a complication could not be applied to spondylitis in this context for it sometimes preceded admissions for treatment of inflammatory bowel disease. The addition of information about death made it possible to define some of the features of ulcerative colitis associated with the subsequent development of cancer of the large bowel and rectum (Nefzger and Acheson 1963). These have subsequently been confirmed in other studies.

In Britain, where in 1962 a single system provided almost all medical care, there seemed to be an opportunity to 'follow the medical record of individuals from cradle to grave' (Acheson, Truelove, and Witts 1961). Thanks to generous financial help from the Nuffield Foundation and subsequently from the Nuffield Provincial Hospitals Trust and the Department of Health and Social Security, the practical challenges of prospective record linkage for a defined community were eventually taken up in Oxford in 1962. The Oxford Record Linkage Study was set four tasks:

(1) To study the feasibility and cost of prospectively accumulating information about key health events, for a defined population, in cumulative personal and family files.

(2) To develop computer methods of record linkage.

(3) To study applications of the files in medical and operational research.

(4) If successful as a pilot study, to promote its extension on a national basis.

Early progress

In the first instance, the data were limited to brief abstracts of each birth, hospital admission, and death for a population of 350 000. The decision not to include any information about treatment other than surgical operations and, therefore, to exclude data about drugs, denied us the opportunity to monitor the unwanted effects of medicines. The exclusion of out-patients limited the value of the file in other ways, for example, in the field of dermatology. On the other hand, inclusion of these categories of data would have substantially increased costs and might have proved overwhelming. It is now clear that two other decisions were crucial in determining subsequent use of the files. The inclusion of data from birth certificates, in addition to data about each pregnancy and delivery, made it possible to link the records of mother and child and thus facilitated a number of studies relating events prior to and during pregnancy to the health of the child in early life. The decision to include information about all deaths in the area, including those outside hospital, enabled the study of survival following discharge from hospital.

Initially we had hoped to link the records into a series of personal cumulative files by using the National Health Service Number but it soon became clear that this number was only available on a minority of records. Fortunately Dr Alex Barr, Statistician to the Oxford Regional Hospital Board (as it was then), had heard Dr Howard Newcombe expound his elegant method of linking records, in spite of omissions and discrepancies in the data, by using names, dates of birth, and other information (Newcombe, Kennedy, Axford, and James 1959). We adopted a system based on similar principles, but modified to the British scene in 1973. This has subsequently been much improved and refined.

The early papers were intended to explore some of the obvious uses of linked files. We were able to demonstrate that even within a single year a substantial proportion of patients (11.8 per cent) are readmitted to hospital (Acheson and Barr 1965). Such readmissions lead to inflation of morbidity rates based on unlinked data. For certain conditions, such as cancer and some psychiatric conditions, morbidity rates based on hospital admissions were found to exceed 30 per cent. The probability of readmission to hospital from home was found to be related not only to the type of illness and to age but also to certain social factors. To begin with, the time-span of the material was too short to disclose associations of aetiological interest between diseases, but even within a single year certain patterns of readmission were obvious. In some conditions, such as cancer, schizophrenia, and diabetes, the chance of being readmitted for the same condition was greater than for all other conditions combined. In another group where the readmission rate was high, the initial diagnosis (for example, 'nervousness', 'abdominal symptoms') had often been tentative, and a more definite diagnosis was reached in the subsequent admission.

We also devised computer methods for deriving hospital morbidity rates based on the person as the denominator rather than the number of admissions (Watts and Acheson 1967). In addition to creating more accurate estimates of morbidity, this made possible the calculation of more accurate hospital fatality rates. We also developed a method of notifying hospitals of deaths which had taken place subsequent to discharge. This revealed that only half of all deaths occurring within one month of discharge had been recorded in the hospital notes (Hubbard and Acheson 1967). A clinical study of a sample of these showed that about one quarter would have been of interest to the clinician in charge of the case (Rang, Acheson, and O'Connor, 1968).

In the first exposition of the value of linked files, Halbert Dunn (1946) emphasized their use in establishing the accuracy or otherwise of recorded data. An early study based on the Oxford file by Alderson and Meade (1967) compared the diagnoses recorded in the hospital records and death certificates of persons who died in hospital. Substantial discrepancies were uncovered which varied with age and type of diagnosis.

The study in Chapter 8, selected from the early papers on reproduction, represents the first attempt to use a linked file to relate events in pregnancy to the subsequent health of the child. Although covering only a single calendar year, it was possible to show the unfavourable effects of low birth weight and low maternal age on the health of the child as indicated by the frequency of hospital admissions in the first year of life. A more detailed study of respiratory illness in early life found striking relationships between respiratory admissions and season of birth (birth during autumn being least favourable), descending birth order, and social class (McCall and Acheson 1968).

Subsequent developments

Following John Baldwin's appointment as Medical Director in 1968, many improvements were made in the quality of the data, the geographical area of coverage was enlarged, and the system of record linkage made more sensitive and specific. Records of psychiatric hospital day-patients and out-patients were included for the first time (Baldwin 1973). Contributions were also made to the statistical analysis of longitudinal files, and a unique set of tables of person-based hospital morbidity rates were made available for a wide range of conditions for the study population (Baldwin and Simmons 1979).

Associations between diseases

From the beginning it had been hoped that analysis of the linked files would disclose relationships between the occurrence of diseases in the same person, thereby giving new insights into aetiology (Acheson, Truelove, and Witts 1961).

Analysis of readmissions, first over a period of five and subsequently over eight years, has revealed an almost bewildering number of statistically significant associations between conditions treated in successive admissions to hospital. It seems that these associations may be divided into three main categories: (1) associations denoting a common aetiology—the associated disorders may occur in any sequence (for example, ankylosing spondylitis and chronic inflammatory bowel disease); (2) sequential associations (including 'complications', like malabsorption following gastrectomy, and long-term iatrogenic effects, for example, of surgical operations or drugs); (3) symptomatic or prodromal associations where an unprecise diagnosis precedes a more exact one (Adelstein, Baldwin, and Fedrick 1979).

A study of the previous and subsequent admissions of 1 329 persons treated for cancer of the large bowel and rectum provides an excellent example of the potential of linked files in this field. A surprise finding was the occurrence of significant excesses of primary cancers of the prostate in men and of the breast in women. However, the analysis gave no support for associations between large bowel cancer and diabetes or coronary heart disease, as would be expected according to the hypothesis of Burkitt and Trowell (1975) concerning dietary fibre. The risk of large bowel cancer in persons previously admitted to hospital for ulcerative colitis (relative risk 25) was clearly seen in this study.

In another study, the records contained in 366 862 personal files, accumulated over a period of eight years, were searched for conditions significantly associated with schizophrenia (see Chapter 7). The pattern of associations was compared with those of samples of patients derived from the same system with affective psychosis and depression. The study provided no support for the suggestion of Katz *et al.* (1967) that schizophrenics are less liable than others to suffer from cancer. One of the most interesting findings was a deficit of cases of rheumatoid arthritis compared to the control groups. This fits with other data suggesting that prostaglandin deficiency may be a factor in the pathogenesis of schizophrenia.

A possibility which was unforeseen at the outset is the use of linked files to focus on areas of particular diagnostic difficulty for medical audit (Hobbs, Fairbairn, Acheson, and Baldwin 1976). For example, in patients treated for large bowel cancer, and for brain tumours, clusters of admissions were identified for diverticulitis and other central nervous conditions respectively in the months immediately preceding admission for the relevant malignant tumour. It seems possible that in some of these conditions the malignant lesion was present but overlooked at the time of the prior admission. The provision of lists of such cases to clinicians for further study might be a useful starting-point for studies of diagnostic techniques.

Maternal and child health

Jean Golding (formerly Fedrick) and Philippa Adelstein have exploited the files to relate events in the early life of the child to those in the relevant pregnancy. It is not possible to do justice to a large number of papers in a few sentences. Of particular interest, however, have been a series of papers about sudden, unexpected death in infancy (Fedrick 1973, 1974a, 1974b). In one of these studies, it was shown that when maternal age and parity are taken into account, artificial feeding does not appear to contribute to risk (Fedrick 1974b). This was important because the prevailing hypothesis at that time was that a large proportion of cot deaths were due to anaphylaxis associated with regurgitation and inhalation of cow's milk protein. The studies pointed to other epidemiological features suggesting the importance of infection as a factor. These features include a higher risk of cot death following delivery in large hospitals compared with small hospitals, clustering of cases in space and time, and a higher than average rate of previous admission to hospital for a respiratory infection in the infants who subsequently died suddenly.

In a study of the epidemiology of infantile pyloric stenosis, the linked files were used not only to relate the occurrence of this condition to factors in the relevant pregnancy, but also to events affecting the general health of the mothers (compared with control women) in preceding years, as indicated by the spectrum of admissions to general hospitals in the two groups (Adelstein and Fedrick 1976). It was also possible to make an estimate of the incidence of pyloric stenosis in sibs of the index child born during the period of the study. This paper illustrates both the flexibility and the potential of linked files in which it is possible not only to relate the experience of mother and child but also of sibs. A study of anencephaly along similar lines is recounted in Chapter 15. An interesting finding, which requires further study, is the association demonstrated between anencephaly and printing as the father's occupation.

Using the Scottish record linkage system (see below) Kendell and his colleagues linked obstetric and psychiatric records for the city of Edinburgh in order to study the epidemiology of puerperal psychosis (see Chapter 9). They showed that a high risk of admission for psychotic illness persists for as long as two years after delivery. Having a first baby, being unmarried, and undergoing a Caesarean section were important risk factors. This study is of particular interest because it links data from two systems of medical records which traditionally have been isolated, both from each other, and from the records of general hospitals.

Drug monitoring

Iatrogenic disease has become a major challenge to preventive medicine, with widespread recognition that exclusive dependence on the alertness of clinicians to detect untoward effects of drugs is inadequate. This is particularly true where there is a long interval between cause and effect or where the clinical features of the side effect are not sufficiently bizarre to attract attention. The difficulty has been how to deal with the enormous scale of drug usage. Skegg and Doll (see Chapter 10) collected data about all drugs prescribed in general practice over a period of two years for a population of almost 50 000 in the Oxford area, and linked them with other medical data concerning the same persons accumulated in the main file of the Oxford Record Linkage Study. Their most interesting finding was a relationship between the prescription of minor tranquillizers and admissions to hospital for road accidents (Skegg, Richards, and Doll 1979). The authors concluded that a similar system on a larger scale would be useful for both generating and testing hypotheses and would be particularly valuable for detecting delayed effects (such as the induction of cancer), sudden deaths outside hospital, and effects of drugs on the foetus. They recommended that a similar system covering at least 500 000 people (but preferably 5 000 000 people) should be set up on a permanent basis.

Material from the Oxford Record Linkage Study has been amplified by data from prescriptions, collected on an *ad hoc* basis by Golding, Vivian, and Baldwin (see Chapter 11), who demonstrated an association between congenital cleft lip and palate and the use of the anti-nauseant Debendox within the first 69 days of pregnancy, but no association with nausea and vomiting. This illustrates another approach to the use of linked records for studying the side-effects of drugs. In this instance, the linked file produced listings of mothers of children with clefts of lip and palate and also facilitated the selection of matched controls. The data about drugs taken in early pregnancy were then obtained by special enquiries to the family doctors of the women concerned. This use of a linked file to test a specific hypothesis does not obviate the additional need for monitoring by means of a system of linked records as recommended by Skegg and Doll.

A national follow-up system

One of the foreseen applications of a system of linked records was the 'provision of accessible follow-up data in respect of death, admission to hospital, etc. for various types of research study'. In fact the framework for such a national follow-up system pre-existed the Oxford Record Linkage Study although it had not been used for this purpose. It consisted of the National Health Service Central Register (NHSCR) which, since 1948, has contained brief particulars of every person registered with a general practitioner, up-dated for deaths and, where known, emigration.

The first known use of the NHSCR for scientific purposes was by Young, Benjamin, and Wallis (1963). In a study of the mortality of widowers, these workers picked a sample of men who were identified on the death certificates of their wives and followed them to death through the NHSCR. A substantial excess mortality (about +40 per cent) was identified in the first six months after widowhood, but it was not clear to what extent a shared unfavourable environment, loss of a caring partner, or physical consequences of grief were responsible.

Subsequently, following the use of the NHSCR in the study of the mortality of asbestos workers by Newhouse (see Chapter 5), the Register has provided follow-up data about death for a large number of epidemiological studies. These have included samples of industrial workers exposed to toxic substances in the workplace, persons who have been subjected to a variety of medical and surgical treatments or who have followed unusual diets, and a variety of other groups. During the 1970s, information about cancer cases from the National Cancer Register were added to the NHSCR so that it is now possible, with appropriate safeguards for confidentiality, for research workers to receive information about living patients with cancer (Office of Population Censuses and Surveys 1981). Information about hospital admissions other than for cancer are not included in the file for England and Wales. The system of record matching used is a manual one based on clerical comparisons of the identifying particulars.

In Canada a generalized iterative record linkage system (GIRLS) has been developed on the basis of the pioneering work of Dr Howard Newcombe. In addition to providing follow-up data for a variety of groups of men and women exposed to agents in the workplace and to drugs, it is intended to study the long-term effects of variations in diet on mortality, by linking data from 20 000 Canadians who participated in the Nutrition Canada Survey conducted between 1970 and 1972 (Smith 1981).

Monitoring for industrial hazards

Although the system described above gives invaluable assistance to research workers testing hypotheses about possible industrial hazards, it cannot in its present form, provide systematic surveillance. This would require data about work histories from at least part of the primary manufacturing sector of industry (Acheson 1981). In the United Kingdom there is doubt that the quality of the data would be sufficient to justify the cost. In the United States, Beebe (1981) has suggested that the one per cent Continuous Work History Sample should be linked with the National Death Index but no comparable sample of work histories is available in the United Kingdom. In Canada, 700 000 individual work histories from a 5–10 per cent sample of the Canadian labour force during the period 1965–71 have been linked to a file of about two million death records and analysis of the study is currently in progress. If the results provide useful insight into the ill-effects of occupational exposure, they may encourage progress along similar lines in the United Kingdom.

Other studies of outcome

Above all, cumulative personal files enable researchers to undertake longitudinal analysis. A number of examples have been quoted above where the end-point of the analysis has been death. In the Oxford system and Scottish system (see below), discharges from hospital for all conditions are recorded and these may also be regarded as 'outcomes' suitable for analysis. This feature of the Oxford Record Linkage Study has recently been used to identify persons at high risk of developing (and dying from) pneumococcal pneumonia, and who might therefore benefit from prophylactic immunization (see Chapter 6). It was found that of 793 persons who had been hospitalized (or died) from pneumonia in the Study population in a single year, 36 per cent who lived and 49 per cent who died had been discharged from hospital in the previous five years. It was estimated that pneumococcal immunization of relatively few discharged patients would prevent each subsequent readmission or death from pneumococcal pneumonia within the next five years. It might therefore be of practical benefit to immunize certain groups of patients against pneumococcal pneumonia on or shortly after discharge from hospital.

The OPCS Longitudinal Study

The OPCS Longitudinal Study, which traces its origin to the Oxford study, is the most important development which has taken place in record linkage in Britain since 1962. It may further prove to be the most significant development in the field of routine social and medical statistics since universal registration of births and deaths became mandatory in 1837 (Office of Population Censuses and Surveys 1973).

The basis of the longitudinal survey is a one per cent sample of the population of England and Wales enumerated at the 1971 Census and selected according to certain dates of birth. The matching of records is carried out clerically. Data about these people from subsequent censuses, the National Cancer Register, and death certificates are linked to the file. The records of persons born on the specified dates after the 1971 Census are added, as are the records of immigrants born on these dates. In effect, the study contains selected records arranged in personal cumulative files for a standing one per cent population sample. Early results have thrown light on the relationships between mortality and a wide range of social factors including household structure, country of birth and parents' countries of birth, migration, occupation, and housing conditions (Office of Population Censuses and Surveys 1978; Fox 1981).

The longitudinal study is limited at present by the fact that a one per cent sample precludes the study of all but the commonest causes of mortality, and, apart from cancer, it contains no morbidity data. However, the early results have illuminated so many problems which had proved intractable to the orthodox cross-sectional approach (for example, the healthy worker effect, the relationships between migration and mortality) that it is already possible to envisage extensions of this study to include additional environmental and morbidity data and a larger sample. The value of this linked file in the study of mortality is illustrated in Chapter 4.

Record linkage in Scotland

Heasman and Clarke (1979) have published a description of the medical record linkage system which exists on a national scale in Scotland. In addition to 100 per cent of the Scottish birth, death, hospital in-patient, and cancer registration records, the system includes school medical examination records and data from the handicapped children's records. The main difference between the Scottish and Oxford systems is that in the former, records are linked on an *ad hoc* basis for particular projects and no permanent system of cumulative personal files is retained.

The Scottish system is being used in both epidemiological and health services research. Under the former heading, studies are being carried out, for example, into the long-term effects of vasectomy and the subsequent health of children who are long-term survivors of cancer. Studies of the development of better indicators of disease incidence using person-based statistics are under consideration. Linkage is also used to calculate total duration of stay where an episode of in-patient treatment is split between hospitals.

Conclusion

The fundamental notion which underlay the setting-up of the Oxford Record Linkage Study was that piecing together data about successive events relevant to the health of individuals would facilitate epidemiological research in various ways. The declared objective, at the time unique, was to prospectively assemble personal cumulative files for a geographically defined population, concerning selected health events and retaining the option to assemble them in family groups. It is a curious reflection of the way ideas develop that it was not until after the Study had been funded that Dr Howard Newcombe's prior work on 'probability matching' of the identification particulars of discrepant and incomplete records was discovered. This led to the adoptionof Halbert Dunn's term 'record linkage' in the title of the study, and to the decision that computer-based 'probability matching' would be used to assemble the files.

Looking back after more than twenty years, I think it fortunate we underestimated the practical difficulties which existed at that time in assembling and handling very large files of data, and in applying to the British scene techniques of record matching developed on more generously identified Canadian vital records. Although questions about what to include in a file of this sort and what to leave out remain as difficult to answer as ever, I personally regret our decision not to include information about drugs. The scale of work could have been limited by, for example, selecting for the cumulative record certain categories of medicine, such as those recently introduced. After all, medicines are chemicals specifically prepared or synthesized because they are known to have an action on the human body. The inadequacies of animal experiments in predicting long-term unwanted effects in man are perhaps more obvious than ever today, and it would seem that the accumulation of properly indexed lists of persons exposed to particular medicines should have high priority.

Others less personally involved than I must evaluate the extent to which the Oxford Record Linkage Study has achieved the first three of its four objectives. Although extension of the Oxford system nationally has not been practicable, there is no doubt that the project has been influential outside the region, for example, in the Scottish system and the OPCS Longitudinal Study. Record linkage is now widely acknowledged as an important epidemiological tool and the term is in general use.

When the study was set up in 1962 the question of whether it was proper for bona fide medical research workers to have access to clinical information from hospitals had barely been asked, although clear rules had long been defined in relation to data from vital certificates and the decennial censuses. As computer technology developed in the 1960s, the propriety of the collation of personal data from different sectors, particularly by government, became a major public issue and the term 'record linkage' began to take on an emotional and political as well as a scientific significance. The future of the work discussed in this volume depends upon the issues surrounding the privacy of records being settled without placing impossible constraints upon epidemiologists. Happily it now appears that in the United Kingdom at any rate such a settlement is in sight.

References

ACHESON, E. D. (1960). Association between ulcerative colitis, regional enteritis and ankylosing spondylitis. *Quarterly Journal of Medicine* **26**, 489.

—— (1967). *Medical record linkage.* Nuffield Provincial Hospitals Trust. Oxford University Press.

—— (1981). Towards a strategy for the detection of industrial carcinogens. *British Journal of Cancer* **44**, 321.

—— and Barr, A. (1965). Multiple spells of in-patient treatment in a calendar year. *British Journal of Preventive and Social Medicine* **19** 182.

——, Truelove, S. C., and Witts, L. J. (1961). National epidemiology. *British Medical Journal* **1**, 668.

ADELSTEIN, P. AND FEDRICK, J. (1976). Pyloric stenosis in the Oxford Record Linkage area. *Journal of Medical Genetics* **13**, 439.

——, Baldwin, J. A., and Fedrick, J. (1979). Cancers of the large bowel: associated disorders in individuals. *Cancer* **43**, 2553.

ALDERSON, M. R. AND MEADE, T. (1967). Accuracy of diagnosis on death certificates compared with that in hospital records. *British Journal of Preventive and Social Medicine* **21**, 22.

BALDWIN, J. A. (1973). Linked record medical information systems. *Proceedings of the Royal Society of London (B)* **184**, 403.

—— and Simmons, H. M. (1979). *Oxford Record Linkage Study: linked record statistical tables* Volume 1: 1963–1970. Oxford University Unit of Clinical Epidemiology.

BEEBE, G. W. (1981). Record linkage and needed improvements in existing data resources. In *Banbury Report 9: Quantification of Occupational Cancer.* (eds. Peto, R. and Schneirderman, M.) pp. 661–73. Cold Spring Harbour Laboratory.

BURKITT, D. P. AND TROWELL, R. C. (1975). *Refined carbohydrate foods and diseases: some implications of dietary fibre.* Academic Press, London.

DUNN, H. L. (1946). Record linkage. *Amer. J. Public Hlth.* **36**, 1412.

FEDRICK, J. (1973). Sudden unexpected death in infants in the Oxford Record Linkage area: I. An analysis with respect of time and place. *British Journal of Preventive and Social Medicine* **27**, 217.

—— (1974a). Sudden unexpected death in infants in the Oxford Record Linkage area: II. The mother. *British Journal of Preventive and Social Medicine* **28**, 93.

—— (1974b). Sudden unexpected death in infants in the Oxford Record Linkage area: III. Details of pregnancy, delivery and abnormality in the infant. *British Journal of Preventive and Social Medicine* **28**, 164.

FOX, J. (1981). Record linkage and occupational mortality. In *Recent Advances in Occupational Health* (ed. J. Corbett McDonald). Livingstone, Edinburgh.

HEASMAN, M. A. AND CLARKE, J. A. (1979). Medical record linkage in Scotland. *Health Bulletin* **37**, 97.

HOBBS, M. S. T., FAIRBAIRN, A. S., ACHESON, E. D., AND BALDWIN, J. A. (1976). The study of disease association from linked records. *British Journal of Preventive and Social Medicine* **30**, 141.

HUBBARD, M. R. AND ACHESON, E. D. (1967). Notification of death occurring after discharge from hospital. *British Medical Journal* **2**, 612.

KATZ, J. KUNOESKY, S., PATTEN, R. E., AND ALLAWAY, N. C. (1967). Cancer mortality among patients in New York mental hospitals. *Cancer* **20**, 2194.

McCALL, M. G. AND ACHESON, E. D. (1968). Respiratory disease in infancy. *Journal of Chronic Diseases* **21**, 349.

NEFZGER, M. D. AND ACHESON, E. D. (1963). Ulcerative colitis in the U.S. Army in 1944: follow-up with particular reference to mortality in cases and controls. *Gut* **4**, 183.

NEWCOMBE, H. B., KENNEDY, J. M., AXFORD, S. J., AND JAMES, A. P. (1959). Automatic linkage of vital records. *Science* **130**, 954.

Office of Population Censuses and Surveys (1973). *Cohort studies: new developments. Studies in Medical and Population Subjects, No. 23.* HMSO, London.

—— (1978). *Household mortality from the longitudinal study. Population Trends, No. 14*. HMSO, London.

—— (1981) *Cancer registration in the 1980s. Report of the Advisory Committee on Cancer Registration. Series M.B.1, No. 6*. HMSO, London.

RANG, E. H., ACHESON, E. D., AND O'CONNOR, B. T. (1968). Clinical significance of deaths after discharge from hospital unrecorded in the hospital notes. *Lancet* **2**, 908.

SKEGG, D. C. G., RICHARDS, S. M., AND DOLL, R. (1979). Minor tranquillisers and road accidents. *British Medical Journal,* **i**, 917.

SMITH, M. E. (1981). Long-term medical follow-up in Canada. In *Banbury Report 9: Quantification of Occupational Cancer* (eds. Peto, R. and Schneiderman, M.) pp. 675–88. Cold Spring Harbour Laboratory.

WATTS, S. P. AND ACHESON, E. D. (1967). Computer method for deriving hospital in-patient morbidity statistics based on the person as a unit. *British Medical Journal* **4**, 476.

YOUNG, M., BENJAMIN, B., AND WALLIS, C. (1963). The mortality of widowers. *Lancet* **2**, 454.

PART I

The theory and methodology of record linkage

1

Record linking: the design of efficient systems for linking records into individual and family histories

H. B. NEWCOMBE

Introduction

In record linking the computer is used as a kind of filing clerk. The task given it is that of building individual and family histories of births, marriages, procreations, deaths, and ill health from the individual registrations of these events, and of doing so on a substantial scale.

Although the computer is at no point asked to carry out any mathematical operation more complicated than simple addition and subtraction, it must nevertheless perform a function that is much more unconventional for machines. It is required to simulate the judgment of a human clerk who attempts to file correctly the incoming correspondence from people who are careless about the way they spell their family names, who may sometimes use their middle names as if these were their first, and who may be writing from places that are not their usual addresses.

Provided that a computer can be instructed to carry out an operation of this kind with a degree of accuracy similar to that of a human filing clerk, the special talent which it may be expected to apply to the task is its speed. Current experience with this sort of computer application is particularly encouraging, in terms of accuracy, speed, and cost, and the capabilities of the machines will undoubtedly increase as time goes on. Thus, it is not unrealistic to think of integrating, in due course, some major fraction of the routine personal documentation dealing with reproduction and health into the form of individual and family histories.

Concepts

A number of concepts will be discussed that are inherently simple, but the implications of these concepts will not necessarily be self-evident.

The idea of linking records, for example, is particularly simple—the phrase

Reprinted from *American Journal of Human Genetics* **19**, 335–59 (1967).

record linking just means bringing together information from two independent sources about the same person—but with successive linkings the information may take on the characteristics of a collection of personal or family histories. Even such familiar file upkeep operations as the insertion of address changes into a mailing list are elementary forms of record linking. However, the process as applied to human genetics will involve successive linkages of routinely collected records of procreative and health events to derive, eventually, multigeneration pedigrees for whole populations.

The two principal steps in any linking operation, namely, those of searching out the potentially linkable pairs of records for detailed comparison and of deciding whether or not a given pair is correctly matched, are commonplace in almost any operation by which a file is kept up to date. However, both of these steps, if they are to be carried out efficiently by machines, involve the use of stratagems of kinds that are employed almost unconsciously by a human filing clerk. For the *searching step*, the aim must be to reduce the number of failures to bring potentially linkable records together for comparison, such as may occur as a result of discrepancies in the file sequencing information, but this must be done without resorting to excessive amounts of additional searching. For the *matching step*, the problem is that of enabling the machine to apply in numerical form the rules of judgment by which a human clerk would decide whether or not a pair of records relates to the same person when some of the identifying information agrees and some disagrees.

Similarly, the idea of arraying pedigree information in linear fashion to facilitate storage, updating, and retrieval by machines using magnetic tapes as the storage medium is simple and by no means new. Nevertheless, the forms which such linear arrays may take bear little resemblance to the conventional pedigree charts with which geneticists are most familiar. The great flexibility of the *linear pedigrees* and the ease with which family relationships of unlimited complexity may be represented in such a fashion are, for this reason, not generally appreciated. In comparison, however, the usual two-dimensional representations are exceedingly cumbersome (Fig. 1.1).

Finally, it has not been uncommon in the past to derive partial histories of individuals and families from the *routine vital and health records*, on a small scale, by manual means. However, the idea that some substantial fraction of these enormous files might be so organized and that we are at the point now where this would be technically feasible and not too expensive is one that has been slow in gaining acceptance. Nevertheless, the inherent possibilities are beginning to be recognized. A colleague of mine is reported to have remarked that we are still using old data on haemophilia, that there are many haemophiliacs in Canada, almost all of whom will wind up in a computer sooner or later, and 'what a shame if it is only opposite a dollar sign'.

The concepts may not be new, but such implications are.

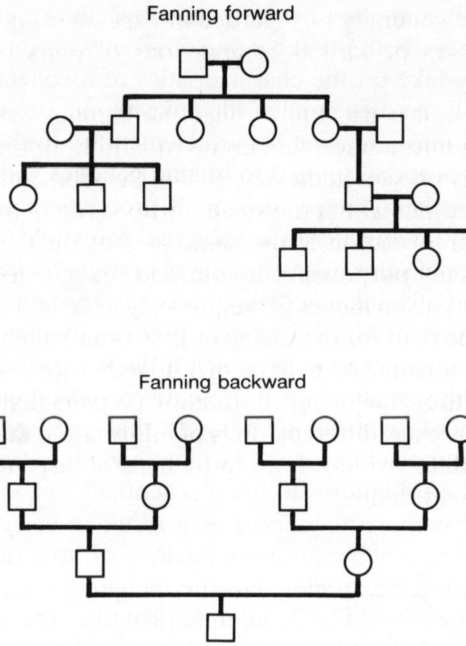

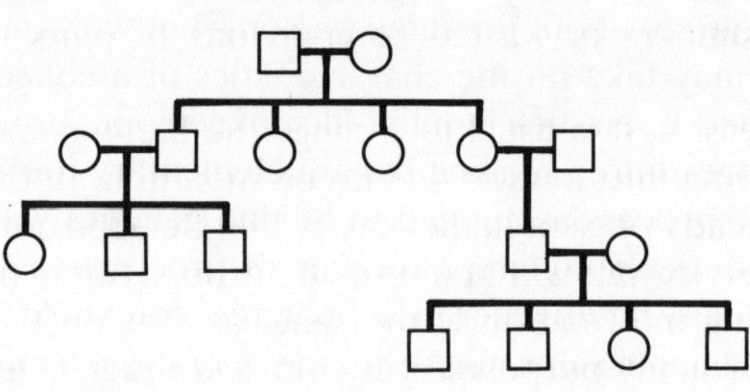

Fig. 1.1 Conventional pedigree charts. Note the difficulty of representing in a single chart the ancestors, descendants, cousins, and in-laws.

Methods of record linking

The two essential steps in the linking of records by computer, that is, the *searching* step and the *matching* step, have precise counterparts in many manual filing operations. Although the accuracies of such operations and the times required are generally regarded as important, it is unusual to judge the efficiencies in numerical terms or to set down the conditions under which an optimum balance may be achieved between the level of accuracy and its cost as indicated by time required to achieve that level. Where such an undertaking is to be carried out on a very large scale by a computer, however, some thought may profitably be given to the efficiency of the operation in these terms.

Optimizing the searching step

In the case of the searching step, errors in the form of failures to bring potentially linkable pairs of records together for comparison could be reduced to zero simply by comparing each incoming record with all of the records already present in the master file. Where the files are large, however, such a procedure would generally be regarded as excessively costly in terms of the enormous numbers of wasted comparisons of pairs of records that are unlinkable.

For this reason, it is usual to arrange the file in some orderly sequence, using identifying information that is common to both the incoming records and those already present in the master file. Detailed comparisons then only need to be carried out within the small portions of the master file for which the sequencing information is the same as that on the incoming records (Fig. 1.2). For many purposes, it is common practice to use the alphabetic surnames and first given names for sequencing a file of personal records. The price that must be paid for the saving of time is an increase in the failures to bring potentially linkable pairs of records together for comparison, owing to discrepancies in the sequencing information on pairs that in fact relate to the same person. However, different kinds of information that might be used for the sequencing differ widely, both in their reliability and in the extents to which they subdivide a file.

Fig. 1.2 Optimizing a single sequence search.

Although alphabetic surnames are commonly employed, they are not particularly efficient for sequencing, because of the high frequency with which they are misspelled or altered. Considerable improvement can be achieved by setting aside temporarily the more fallible or labile parts of the information which the surnames contain, while retaining as much as possible of the inherent discriminating power. There are a number of systems for doing this, the most common of which is known as the Russell Soundex code. This is essentially a phonetic coding, based on the assignment of code digits which are the same for any of a phonetically similar group of consonants.

In practice, we have found that the Soundex code remains unchanged with about two-thirds of the spelling variations observed in linked pairs of vital records, and that it sets aside only a small part of the total discriminating power of the full alphabetic surname. The system is designed primarily for Caucasian surnames, but works well for files containing names of many different origins (such as those appearing on the records of the US Immigration and Naturalization Service). This particular code is less satisfactory, however, where the files contain names of predominantly Oriental origin, because much of the discriminating power of these resides in the vowel sounds, which the code ignores.

Any kind of identifying information that is available on all of the records may, of course, be used for sequencing the files, and it should not be assumed that surnames necessarily possess special merit for this purpose. The qualities required are reliability and discriminating power, both of which may be measured numerically. Usually, where the discriminating power of any one kind of information alone is insufficient to divide the file finely enough, two or more kinds of information may be used together to achieve a required degree of subdivision. However, each additional kind of information carries its own likelihood of discrepancy and thus contributes to the overall tendency for the sequencing information to be reported differently on successive records relating to the same person, with a resulting increase in the frequency with which potentially linkable records will fail to be brought together for comparison. It is important, therefore, to choose the most appropriate kinds of information from among those that are available.

Fortunately, there are numerical tests which will indicate the relative merits of the different items of identifying information for the purpose of sequencing the files. Three values will be discussed, the *coefficient of specificity*; the *discriminating power*, which is simply another way of describing the specifity; and a so-called *merit ratio*, which may be used to indicate the amount of discriminating power per unit likelihood of discrepancy. This latter value can be used in selecting the most appropriate information to be employed in sequencing a file.

The fineness with which a file will be divided by a particular kind of identifying information may be represented by a single number, the *coefficient of specificity*,

$$C_s = \Sigma P_x{}^2 \tag{1.1}$$

where P_x is the fraction of the file falling in the xth block (see Fig. 1.3). C_s may be thought of as the fraction of the file falling within a block of strictly representative size. Since most identifying information divides a file unevenly into a mixture of small and large blocks, it is convenient to be able to indicate the effective degree of division of the file in this simple manner.

Unlike the coefficient of specificity, which gets smaller as a file becomes more finely divided, the *discriminating power* increases with the extent of the subdivision. Furthermore, it is usually regarded as an 'addable' quantity. Thus, the discriminating power may be taken as the logarithm of the inverse of the coefficient of specificity, and in practice we have found it convenient to use logarithms to the base two (see Table 1.1):

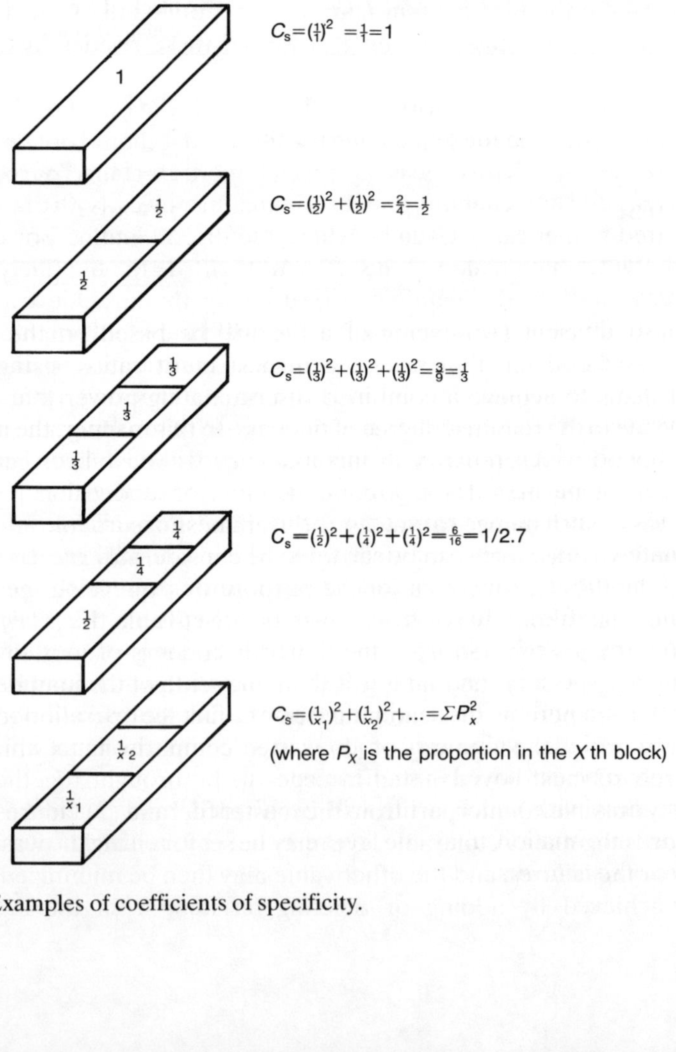

$$C_s = \left(\tfrac{1}{1}\right)^2 = \tfrac{1}{1} = 1$$

$$C_s = \left(\tfrac{1}{2}\right)^2 + \left(\tfrac{1}{2}\right)^2 = \tfrac{2}{4} = \tfrac{1}{2}$$

$$C_s = \left(\tfrac{1}{3}\right)^2 + \left(\tfrac{1}{3}\right)^2 + \left(\tfrac{1}{3}\right)^2 = \tfrac{3}{9} = \tfrac{1}{3}$$

$$C_s = \left(\tfrac{1}{2}\right)^2 + \left(\tfrac{1}{4}\right)^2 + \left(\tfrac{1}{4}\right)^2 = \tfrac{6}{16} = 1/2.7$$

$$C_s = \left(\tfrac{1}{x_1}\right)^2 + \left(\tfrac{1}{x_2}\right)^2 + \ldots = \Sigma P_x^2$$

(where P_x is the proportion in the Xth block)

Fig. 1.3 Examples of coefficients of specificity.

$$D_p = \log_2(1/C_s) \tag{1.2}$$

Finally, the merit of any particular kind of identifying information for sequencing the files may be taken as the ratio of the discriminating power to the likelihood of discrepancy or inconsistency of such information in linkable pairs of records:

$$M_t = D_p/I \tag{1.3}$$

In calculating this so-called *merit ratio*, we normally use the percentage likelihood of inconsistency as the numerical value of I.

Table 1.1 Relationship of coefficient of specificity and discriminating power

Coefficient of specificity $C_s = \Sigma P_x^2$	Discriminating power $\log_2 (1/C_s)$	Equivalent number of blocks if file equally divided
1	0	$2^0 = 1$
1/2	1	$2^1 = 2$
1/4	2	$2^2 = 4$
1/8	3	$2^3 = 8$
1/16	4	$2^4 = 16$
1/1024	10	$2^{10} = 1024$
$1/10^6$	20	$2^{20} = 10^6$

The most efficient sequencing of a file will be based on the items of identifying information that have the highest merit ratios, using enough different items to achieve a combined discriminating power that will subdivide the file to the required degree of fineness. In this manner, the minimum total likelihood of discrepancy or inconsistency will have been introduced into the sequencing items for any required degree of subdivision.

By means of such numerical values, the usefulness of surname information in its Soundex-coded form can be shown to be considerably greater than that of the full alphabetic surnames for the purpose of sequencing the files, the merit ratio being about two or three times as large (Table. 1.2). The residual information that is omitted from the Soundex codes is of very low quality indeed, having a merit ratio that is less than one-tenth of the Soundex codes.

The approach permits the searching step of a linkage operation to be optimized, in terms of the numbers of (1) wasted comparisons to which an incoming record must be subjected in order to be brought together with a potentially linkable counterpart from the master file, and (2) failures to bring such records together. A tolerable level may be set for either the wasted comparisons or the failures, and the other value may then be minimized. Adjustment is achieved by adding or deleting an item from the sequencing

Table 1.2 Relative merits of alphabetic versus Soundex coded surnames for sequencing files

Surname information	Discriminating power D_p	Equivalent number of blocks of equal size $1/C_s$	Percentage likelihood of discrepancy* I	Merit ratio $Mt = D_p/I$
Alphabetic	+9	512	2.2	4.1
Soundex	+8	256	0.8	10.0
Residual	+1	2	1.4	0.7

* Average for husbands' and wives' birth surnames.

information, thus increasing or decreasing the fineness of subdivision and the errors simultaneously until the required balance is struck. At no time should the sequencing information include an item with a lower merit ratio where one with a higher ratio is available. The cost of the searching step is thus balanced against its precision with a view to getting the best possible bargain.

In practice, we have found that by sequencing a master file of 114 000 marriage records in order of the pairs of surname codes for the grooms and brides, the number of wasted comparisons was kept at a very low level, i.e. 0.6 per incoming birth record where the births had arisen from marriages represented in the master file and 1.6 for all other incoming birth records. The number of failures to bring potentially linkable records together for comparison due to spelling discrepancies that altered one or other of the Soundex codes amounted to 1.6 per cent of the potentially possible linkages.

The discussion so far has assumed that all of the linkings will be carried out using files arranged in a single sequence. However, the cost of sorting by computer is rapidly diminishing. Where more than one sequence is permitted, an even better bargain may be struck in terms of the precision that can be achieved for any given number of wasted comparisons. Linkings may then be carried out using very fine subdivisions of the file sequences, based on information of quite limited reliability, with the assurance that potentially linkable pairs of records which are not brought together on the first search will be compared in one of the alternative sequences based on other identifying information.

One quite large manual test of such a procedure has been carried out in which initials and provinces of birth were substituted in the secondary sequences for one or other of the two surname codes. This test showed that a reduction in errors by more than tenfold could be achieved at the price of a two- to threefold increase in wasted comparisons.

Where the avoidance of 'lost' linkages is of special importance, the use of multiple alternative sequences represents an ultimate in refinement.

Optimizing the matching step

When pairs of records are brought together for comparison, decisions must be made as to whether these are to be regarded as linked, not linked, or possibly linked, depending upon the various agreements and disagreements of items of identifying information. It is also desirable that such decisions be based on numerical estimates of the degrees of assurance that the records do or do not relate to the same persons. The computer is asked, in effect, to simulate the processes of human judgment and to make the best use it can of the items of identifying information that are individually unreliable but collectively of considerable discriminating power.

The extent of the personal information that is usually entered in the vital registration makes the potential accuracy of the linkings of these records high indeed. New-born children, grooms and brides, and deceased persons are commonly identified by their full birth names, their birth dates or ages, and their birthplaces. Together with this personal identification, there is a substantial amount of family information. The full names of the parents, including the maiden surname of the mother, are usually given, as well as their birthplaces. In addition, the ages of married couples are entered in the records of their marriages and the records of the births of their children (Table. 1.3).

Thus, there is an abundance of overlapping information that may be used to link (1) deaths to births, (2) births to the parental marriages and to the

Table 1.3 Identifying information on vital records

Event and individual	Birth name	Birthplace*	Birth date (or age)
Marriage			
Groom	+	+	(+)
Bride	+	+	(+)
Father of groom	+	+	
Mother of groom	+	+	
Father of bride	+	+	
Mother of bride	+	+	
Birth			
Child	+	+	+
Father	+	+	(+)
Mother	+	+	(+)
Death			
Deceased	+	+	+
Spouse	+		
Father	+	+	
Mother	+	+	

* i.e. city or place, and province or country.

births of older siblings, and (3) marriage records of brides and grooms to their birth records, to the marriage records of their parents, and to the birth and marriage records of their siblings (Table 1.4). Even where some of the items fail to agree, the combined discriminating power of such information is almost always large.

A human filing clerk attempting to carry out such a grouping operation would intuitively attach greater positive weight to some of the agreements than to others and greater negative weight to some of the disagreements than to others. In each instance, the question that is asked, almost unconsciously, is 'Would such an agreement be likely to have occurred by chance if the pair of records *did not* relate to the same person?' or 'Would such a disagreement

Table 1.4 Examples of kinds of linkage

Event		Parental information (husband × wife)				Individual information	
Kind	Year	Surnames	Initials	Birthplace codes	Ages	Name	Birthdate (or age)
Death to birth							
Birth	1950	Doe × Cox	JA	MB 09 09	30 25	Fred	15.6.50
Death	1955	Doe × Cox	JA	MB 09 09	— —	Fred	15.6.50
Birth to parental marriage							
Parental marriage	1945	Doe × Cox	JA	MB 09 09	25 20	—	—
Birth	1950	Doe × Cox	JA	MB 09 09	30 25	Fred	15.6.50
Marriage of a groom, to own birth and own parents' marriage							
Parental marriage	1945	Doe × Cox	JA	MB 09 09	25 20	—	—
Birth	1946	Doe × Cox	JA	MB 09 09	26 21	Andy	18.5.46
Own marriage	1966	Doe × Cox	JA	MB 09 09	— —	Andy	(age 20)

be likely to have occurred by chance if the pair of records *did* in fact relate to the same person?' The answer in each case will depend upon prior knowledge gained from experience. An initial known to be rare, such as 'Z', will be regarded as less likely to agree by chance on a pair of records than would a commonly occurring initial, such as 'J'. Similarly, a highly reliable and stable item of identification, such as sex, when it fails to agree, will argue more strongly that the people referred to are *not* the same than would, for example, disagreement of province of birth, which is known from our own experience to be discordant in about one per cent of genuinely linked pairs of records.

The mathematical basis of such intuitive assessments is really quite simple. In general, agreements of initials, birth dates, and such will be more common in genuinely linked pairs of records than in pairs brought together for comparison and rejected as unlinkable. The greater the ratio of these two frequencies, the greater will be the weight attached to the particular kind of agreement.

If we wish to obtain numerical weights that can be added to other such weights, the above ratio may simply be converted to a logarithm. In practice, the logarithm to the base two has proved particularly convenient. These so-called *binit weights* are simply

$$W_t = \log_2(A/B) \tag{1.4}$$

where A and B are the frequencies of the particular agreement, defined as specifically as one wishes, among linked pairs of records and among pairs that are rejected as unlinkable. The binit weights for agreements will have positive values because A in such circumstances is always greater than B (Fig. 1.4), and these weights may be regarded as strictly analogous to the discriminating powers discussed earlier except that they relate to particular values of the various items of identifying information.

There is no need to alter this formula when deriving the weights for disagreements. A and B may be regarded simply as the frequencies of the particular disagreement, defined in any way, among linked and unlinked pairs of records. Usually the weights will then be negative in sign, because disagreements are, in most instances, less common among the linked than among the unlinked pairs, i.e. A will be less than B and the logarithm of A/B will be negative.

Exceptions will occur in which an apparent disagreement is in reality a partial agreement. For example, a discrepancy of one year of age, after allowance is made for the interval of time between the two registered events, will

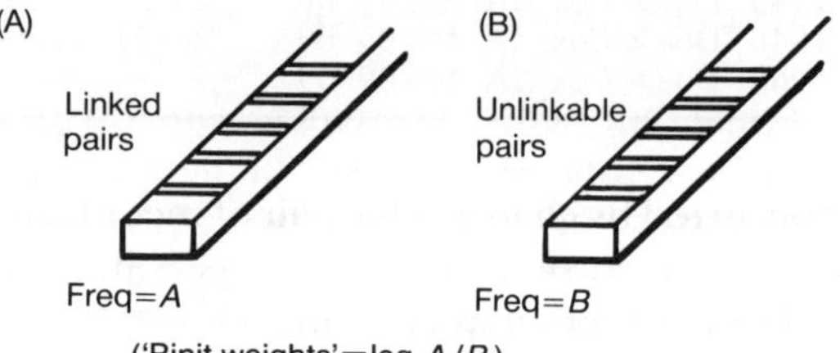

Examples

Kinds of agreements or disagreements	Frequency in linked pairs A	Frequency in unlinkable pairs B	Ratio A/B	Binit weight $\log_2$ A/B
Agreements				
Male sex	1/2	1/4	2	+1
Initial 'J'	1/16	1/256	16	+4
Initial 'Z'	1/1 000	1/1 000 000	1 000	+10
Disagreements				
City of residence	1/3	2/3	1/2	−1
Initial (any)	1/40	32/40	1/32	−5
Sex	1/8 000	1/2	1/4 000	−12

Fig. 1.4 Calculating 'binit' weights.

frequently be a reflection of an underlying genuine agreement. Fortunately, however, it is not necessary to prejudge the issue. If the apparent discrepancy is predominantly a reflection of a partial agreement, the calculated weight will automatically turn out to be positive.

In practice, the formula is used to derive from the actual files a set of look-up tables of weights for agreements and disagreements of various items of information, broken down by the natures of these agreements and disagreements to whatever extent is necessary to make nearly full use of the discriminating powers. Such tables are stored in the memory of the computer. For each detailed comparison of a pair of records, the positive and negative weights appropriate for the different agreements and disagreements are added together, and the total weight is used to indicate the degree of assurance that the pair do, or do not, relate to the same person. The procedure assumes as a tolerable approximation that the weights for the individual agreements or disagreements are uncorrelated with each other; corrections are possible where this is not strictly true, but in our own experience these have been too small to be worth applying.

The derivation and use of the binit weighting factors have been described in greater detail elsewhere (Newcombe *et al*. 1959; Newcombe and Kennedy 1962). For present purposes, it is sufficient to indicate that there is great flexibility in the manner in which the weights can be employed and that they permit the introduction of numerous refinements so as to make nearly full use of the discriminating power inherent in the identifying information. For anyone planning an actual application, I would recommend that a number of small linking studies be carried out by hand to provide an opportunity to experiment with the system and become familiar with its characteristics.

The total binit weight represents the extent to which assurance of a genuine linkage is increased, or decreased, as a result of the comparisons made. Such weights are, in fact, logarithms to the base two of the factors by which the odds in favour of a linkage are increased over and above what they would have been in the absence of the comparisons.

In our own operation, the linkages are carried out within the very small 'double surname pockets' of the master file, which contain on the average between one and two records apiece. Furthermore, an incoming record is quite likely to find a linkable counterpart there. Thus, even in the absence of the detailed comparisons, the probability of a match with a record drawn at random from the correct pocket of the master file will not be so very much less than 50 per cent (i.e. odds of $1:1$). In this situation, the total binit weight will closely approximate the $\log_2$ of the odds in favour of a linkage. Weights of $+10$ and of $+20$, for example, may in this situation be regarded as indicating favourable odds of approximately 1 000 to 1 and 1 000 000 to 1 respectively.

Using the double-surname sequenced files in this manner, no weights are attached to agreements of the items of sequencing information, i.e. to agreements of the surname codes. The reason is that the discriminating powers of these have already been taken into account automatically, since it is this information which determines the sizes of the pockets in the master file.

If binit weights were attached to agreements and disagreements of the sequencing information, incoming records would then have to be thought of as linking within a population of records consisting of the whole of the master file. Suppose, for example, that this contained 10^6 records and was known to include one which matched each of the incoming records. Under these conditions, the chance of an incoming record linking with a randomly chosen record from the master file would be $1/10^6$ $(= 2^{-20})$. However, if the detailed comparisons yielded a weight of $+24$, this would raise the odds from 2^{-20} up to 2^4, i.e. to $16:1$ in favour of a genuine linkage.

Thus, to derive from the total binit weights the odds in favour of a linkage, allowance must be made for the size of the population of records within which the linkage is carried out by subtracting $\log_2$ of this population size. Similarly, allowance must also be made for the limited probability that there is, in fact, a matching record within that particular population. The $\log_2$ of this probability will be negative in sign and when added to the total binit weight will further reduce its value.

In practice, thresholds must be set which specify the ranges of binit weights which are to be regarded as representing linkage, no linkage, and possible linkage. Initially, these thresholds may be set to what seem intuitively to be reasonable values, but empirical tests are needed to ensure that false linkages, failures to link, and tentative linkages are balanced in a reasonable fashion.

In an actual operation, the total weights for linked pairs should be recorded permanently as evidence of the degree of assurance on which the linking was based. Similarly, for pairs of records that are judged to be neither positively linkable nor positively nonlinkable but which represent the most likely linkage available, it is prudent to retain permanently information about each such doubtful link and the weight associated with it. As more information accumulates about the family groupings, such as the sequences of birth orders in the families and the intervals between the births, this further knowledge may assist with the resolution of some of these doubtful linkings, provided that the information about them is retained on the files.

Factors affecting the speed of the record linking operation

A number of practical considerations will influence the speed of a record linking operation.

The individual magnetic tape records should not be unnecessarily large, as this will increase the times required for input and output and for sorting the records. It will also limit the number of records that can be manipulated within the available core memory at any one time. The record format chosen for our own linking operation, using the vital registrations, consists of 25 words of 30 or 32 bits each (depending upon the magnetic tape units used). Each word may contain ten octal digits or five alphanumeric characters. This size of record was found to be sufficient for the storage of the individual and family identifying information, the statistics, and the cross-referencing information pertaining to a vital registration (Table. 1.5).

Speeds are also affected by the amount of unused space on the magnetic tapes between records or between 'blocks' of records. On the tapes used with the Control Data G20 computer, on which most of the recent work was done, records are stored in addressable blocks of 800 words each, i.e. containing 32 records per block. If records are read singly on to tape rather than in blocks, a substantial fraction of the tape is used up in the inter-record gaps.

Table 1.5 Typical magnetic tape format for a vital record

Information	Word*
Soundex pair	1
List word	2
Event (date, etc.)	3–6
Husband (name, etc.)	7–9
Wife	10–12
Offspring	13–14
Record linkage cross reference	15–17
Sibship cross reference	18–19
Statistics	20–24
Other cross reference	25

* One word equals ten octal digits or five alphanumeric characters.

A special time-saving feature in our own linking operation has been the use of the so-called 'list processing' method. Records entering a husband-wife double surname pocket in the master file are arranged, physically, simply in order of their entry or acquisition, regardless of the appropriate logical sequence in the family groups. The logical position of each record is indicated by the inclusion on it of the 'entry number' (i.e. acquisition number) of the record that logically precedes it and that of the record that logically succeeds it. These numbers are known respectively as the backward and forward links.

When a new record enters the double surname pocket, known as a 'super-family', it is placed physically at the end; backward and forward links are then entered in the incoming record, and the existing links on the records that immediately precede and succeed it in the logical sequences are updated (Table 1.6). The saving of time occurs because with this procedure there is no need to alter the physical positions of the records already in a pocket to make room for a new record each time one is to be interfiled. The list processing method used has been described in detail by Kennedy *et al*. (1964).

Table 1.6 Example of list processing

New record	Position	Record	Links	
			Forward	Back
G	(1)	G*	0	0
B	(1)	G	0	2
	(2)	B*	1	0
D	(1)	G	0	3
	(2)	B*	3	0
	(3)	D	1	2
F	(1)	G	0	4
	(2)	B*	3	0
	(3)	D	4	2
	(4)	F	1	3
A	(1)	G	0	4
	(2)	B	3	5
	(3)	D	4	2
	(4)	F	1	3
	(5)	A*	2	0

* Indicates 'flag' for head of list.

Another factor that affects the speed of a linking operation has been mentioned earlier, namely, the size of the units into which the file is broken by the sequencing information. In our own experience, the use of two phonetically coded surnames relating to the husband-wife pair has divided a master file of 114 000 marriage records into units containing on the average about 1.6 records each. For approximately 80 per cent of the file the pairs of surname codes are unique, i.e. they occur only once in that combination throughout the whole file.

Under the various conditions described above and pertaining to our own operation, incoming birth records have been merged and linked with a master file of parental marriages and earlier births at a rate of 2 300 per minute. Thus for the British Columbia population of 1.6 million people, with which this study is concerned, a year's crop of 35 000 birth records can be merged and linked with the master family file of ten years of marriages in somewhat less than 30 minutes of machine time, once the magnetic tape records have been prepared in the proper format and appropriately sequenced. At a machine rental of two dollars per minute this is equivalent to a cost of 0.1 cents per record, i.e. it is minute in comparison with the cost of producing the punchcards in the first place, as is done routinely for administrative and statistical purposes.

The ways in which these various time-saving devices have been employed are described in greater detail by Kennedy *et al.* (1965).

Storage and retrieval

In the sections that follow, we will consider the manner in which records relating to sibship groups may be stored together, certain extensions of the procedures to permit the inclusion of pedigree information covering an indefinite number of generations, and methods of retrieving information from the sibship grouping and multigeneration pedigrees. The records pertaining to the sibships, of course, fall within the main file sequence based on the surname pairs in their phonetically coded forms (Table. 1.7).

Table 1.7 Example of double Soundex file sequence*

Adams × Adair	A 352	A 360
Adams × Baron	A 352	B 650
Adams × Caird	A 352	C 630
Adams × Danys	A 352	D 520
Baker × Allen	B 260	A 450
Baker × Barks	B 260	B 620
Baker × Caron	B 260	C 650
Baker × Duffy	B 260	D 200
Baird × Aubry	B 630	A 160
Baird × Baker	B 630	B 260
(and so on)		

* i.e. by husband's surname code followed by the wife's maiden surname code.

Storage of sibship groupings of records

There is a natural sequence in which the vital and health records pertaining to a sibship group may be linked and stored. Starting with the parental marriage registration, which may be regarded as a 'head-of-family' record, birth records are linked to the marriage record in chronological order, and records of the various events of ill health, including death, are linked to the birth records of the children to whom they relate, those for a particular child falling likewise in chronological order after his or her birth record (Table 1.8).

The experience which we have had with this kind of file organization relates to records of marriages, livebirths, stillbirths, and deaths, together with those from a special register of handicapping conditions of children and adults. In addition, detailed plans have been worked out for the possible future inclusion of substantial numbers of records from a universal scheme of hospital insurance. Off-line linkings with the birth registration records are needed in the case of the handicap and hospital records in order to pick up

Table 1.8   Example of a sibship group of records

	Parental couple	Child
Parental marriage	Doe × Cox	—
Birth 1	Doe × Cox	Alan
Birth 2	Doe × Cox	Carl
Ill health	Doe × Cox	Carl
Death	Doe × Cox	Carl
Birth 3	Doe × Cox	Edna

the mother's maiden name which is lacking on the original form. Only after this has been done can the handicap and hospital records be merged and linked with the master family file, which is arranged in order of the two parental surname codes.

Incompleteness of a sibship grouping of records poses no special problem. In the absence of the parental marriage record, for example, the birth record of the oldest child represented in the file may serve as the head-of-family record, and records of the births of younger siblings will be linked to it. A death record may serve likewise as a head-of-family record where it relates to the oldest child represented in the family group and the birth record for this child is missing. Thus, all of the available records of vital and health events may be merged and linked into sibship arrays, regardless of the degree of completeness or incompleteness of these groupings, and the master file may be updated periodically by the introduction into it of successive crops of current records.

The times required to merge and link the death and handicap records to the master file are somewhat greater than those for the corresponding operation as applied to birth records. There are two reasons for this. First, an ill health or death record must scan all of the birth records present in the appropriate double surname pocket of the master file, and these will tend to be more numerous than the head-of-family records which the incoming births must scan. Second, where an incoming ill health or death record fails to find a matching birth record, it must scan the double surname pocket a second time in an attempt to find a head-of-family with which to link.

In our own operation, handicap and death records were merged and linked with the master file at a rate of approximately 1 000 per minute, i.e. at about one-half of the speed for the merging and linking of birth records.

Storage of multigeneration pedigrees

The modifications of the above procedures needed to permit the linking and storage of the vital and health records in the form of multigeneration pedigrees are surprisingly simple. For most registration areas, the marriage records contain sufficient information to serve as bridges between the generations and between the in-law sibships.

Information from a marriage record may be treated in two ways. We have discussed already how it can be arranged into the form of a head-of-family record representing the marriage of a parental couple. Similarly, information from the registration form may also be fitted into the format of a record such as is used to describe an event in the life of an individual. The part of this latter kind of record entry that is assigned to family information would then contain the names and other identifying particulars of the parents of the newly married person, and the part of the record assigned to personal identification would contain his or her own name, age, and birthplace. This kind of entry of the marriage information is almost precisely analogous to a death record, since both relate to events in the lives of members of a sibship group. In the master file, the three entries pertaining to a particular event of marriage (i.e. the groom's entry, the bride's entry, and the head-of-family entry) will each become part of a different sibship group of records.

The only special requirement for the three marriage entry records is that each of them, before being placed in these various locations on the master tape, be cross-referenced to the other two. This is done by inserting in the cross-reference field of each record entry the double surname codes for the other two. These codes, together with the marriage registration number which is common to all three entries, provide both a means of access within the master file from one of the double surname pockets to the other two and a positive identification of the alternative entries when the pockets in which they occur have been located. The cross-referencing is illustrated in Tables 1.9 and 1.10.

The simplicity of the procedure resides in the use of essentially the same format for the marriage entries of grooms or brides as for their death records. In our own operation, the same programs that are used to build the sibship groupings of records will also be employed to insert into these groupings the grooms' and brides' marriage entries, just as they would the records of any other kinds of events in the lives of the same individuals.

The idea of thus putting family groups of records into a single linear array and of using cross references to indicate the relationships between the groupings that are filed as units is basic to any system by which computers may be employed to store and retrieve large quantities of pedigree information of unlimited complexity. The special features of the system described are merely matters of convenience. The choice of the sibship group as the unit of storage and of the surname pair as the sequencing information may have fairly wide application, but the details of the use of identifying particulars have been dictated largely by the nature of the vital records.

It would, of course, be feasible to store the same pedigree information more compactly if the family relationships were worked out in advance so that every individual could be assigned an identifying number containing as

Table 1.9 Example of a marriage registration and of the marriage entry records derived from it

Marriage registration

Groom	Dunn, Alex
Bride	Rowe, Anna
Groom's father	Dunn, Carl
Groom's mother	Bell, Edna
Bride's father	Rowe, Paul
Bride's mother	Hill, Jean

Marriage entry records

	Parental couple	Offspring
1. Head of family entry	Dunn × Rowe (Alex) (Anna)	—
2. Groom's entry	Dunn × Bell (Carl) (Edna)	Alex
3. Bride's entry	Rowe × Hill (Paul) (Jean)	Anna

Table 1.10 Example of cross-referencing a sibship to the related sibships

Record	Parental couple	Offspring	Cross references
Parental marriage	Dunn × Bell		Dunn × Nash—father's sibship Bell × Mann—mother's sibship
Birth 1	Dunn × Bell	Alex	
Groom's entry	Dunn × Bell	Alex	Dunn × Rowe—new family Rowe × Hill—bride's sibship
Birth 2	Dunn × Bell	Stan	
Groom's entry	Dunn × Bell	Stan	Dunn × Knox—new family Knox × Flynn—bride's sibship

few digits as possible, but the disadvantages of this approach where large populations are involved should perhaps be mentioned. A main objective of the present handling procedures has been to avoid entirely all manual manipulations so that full use can be made of the speeds of electronic computers. If this feature is to be preserved, the present kind of linking operation would have to be carried out anyway. A more important problem would be what to do with the borderline linkings when condensing the pedigree information into its more compact form, since both the extents of the uncertainties and the means for their later resolution would tend to be lost in the process. It might also be difficult to keep open the possibility, as the present system does, of merging at some future time the pedigrees drawn from a limited region, such as a province or a state, with those for a wider region such as the country as a whole.

Retrieval of pedigree information

The need for writing detailed programs does not end with the establishment of a master family file containing the required pedigree information. For almost any kind of genetic study, the extraction of the required tabular information from a printed listing of the master file would be almost unthinkably laborious and expensive.

In general, it is necessary first to prepare programs that will summarize in a single record whatever information is required about a particular family. A further program is then written to extract information in tabular form from the resulting file of these summary records. Two examples of such procedures will be described, relating to sibship groups and to multigeneration pedigrees respectively.

Where the family units under study are restricted to the sibships, summaries of the events of birth, ill health, and death in the lives of the various members of a sibship will usually be derived in two steps. First, individual histories will be condensed so that there is just a single summary record for each child replacing the separate records for the various events. The resulting magnetic tape file of individual or personal summaries can be used repeatedly to prepare the much more compact family summary records, which may be of a variety of kinds, depending upon the natures of the studies for which they are to be used (Table 1.11).

To facilitate subsequent tabulations, the family summary records will have a different fixed field for each of the siblings. There must also be provision for large families, which will sometimes overrun a family summary record of modest size. This is best taken care of by arranging for trailing records to act as extensions where needed.

In one study which we have done using this procedure the coded causes of stillbirths, handicaps, and deaths were entered into the fields of the family summary record assigned to the particular siblings who were affected; and for the unaffected siblings just the fact of birth, the birth order, and the sex of the child were entered.

In this particular study, use was made of the family summaries to derive information about the magnitudes of the risks to the later-born siblings of children who had been stillborn, handicapped, or had died, as the result of diseases of various kinds. The tabulations contained, typically, the number of index cases of a disease, the numbers of earlier and later siblings of the index cases, and the number of later-born siblings suffering from the same condition (Table 1.12). For details of the computer programs by which the

Table 1.11 Example of individual and family summary records

Event records for a sibship (one per event)

Event code	Birth order	Family	Child	Disease code
J (birth)	1	Fox × Dow	Alan	—
J (birth)	2	Fox × Dow	John	—
J (birth)	3	Fox × Dow	Vera	—
Q (handicap)		Fox × Dow	Vera	123
J (birth)	4	Fox × Dow	Leon	—
R (death)		Fox × Dow	Leon	456

Individual summary records (one per child)

(J)	1	Fox × Dow	Alan	—
(J)	2	Fox × Dow	John	—
(Q)	3	Fox × Dow	Vera	123
(R)	4	Fox × Dow	Leon	456

Family summary record (one per sibship)

(Fox × Dow)	1 (J) ———,	2 (J) ———,	3 (Q) 123,	4 (R) 456

different steps in the extraction were carried out, the reader is referred to Smith *et al.* (1965).

A more elaborate procedure is required where multigeneration pedigrees are to be summarized, because as an initial step the sibship groupings of records relating to a particular family must be brought together from different parts of the master file. Before starting this step, certain sibships whose relatives one wishes to ascertain will have been extracted from the master file. These may be called 'index sibships', and they will in most instances have been chosen because they include individuals who are affected by some disease of special interest.

The records of the index sibships may contain cross-referencing information (in the form of double-surname codings and marriage registration

Table 1.12 Example of a tabulation from family summary records

Disease code 325 (mental deficiency)

	Normal	Stillborn	Handicapped	Dead	Handicapped and dead
	(J)	(K)	(Q)	(R)	(S)
Index cases	0	0	506	9	58
Earlier sibs	208	2	6	16	0
Later sibs, same cause	0	0	11	0	1
Other later sibs	286	2	11	14	0

numbers) indicating links with as many as six different kinds of related sibships, i.e.

(1) From the parental marriage (head-of-family) records to:
   (a) the fathers' sibships and
   (b) the mothers' sibships.

(2) From the marriage records of the 'affected' individuals who got married (i.e. from the grooms' and brides' entries) to:
   (c) their offspring's sibships and
   (d) their spouses' sibships.

(3) From the marriage records of the brothers and sisters who got married to:
   (e) the sibships of the nephews and nieces of the affected individuals and
   (f) the sibships of the spouses of the brothers and sisters who got married.

These six different kinds of cross references may be used in a single scan to draw from the master family file all of the groups of records pertaining to sibships that are removed by one degree of relationships from those in which the affected individuals occurred, including the in-law groups (Fig. 1.5).

Similarly, in a second scan of the master tape, use may be made of the further cross-referencing information contained in the sibship groups of these six different kinds to extract the sibships that are removed by two degrees of relationship from those in which the affected individuals occurred. Again, the in-law sibships may be extracted in the same way as those of the blood relatives. And so, with each successive scan, an expanding circle of more distant relatives may be identified and retrieved from the master file.

Each such scan will be exceedingly rapid, even where large numbers of sibships groups are extracted. Thus, it is feasible to carry out the retrieval of multigeneration pedigrees on a truly massive scale.

From this point on, the making of summaries would follow much the same pattern as described earlier, except that the family summary record might be more complex than the sibship summary record.

The chief limiting factor in work of this kind is not the speed of the computer but the time required to develop the appropriate programs.

The likelihood of future 'total utilization' of pedigree information

Geneticists will at first tend to think of the possible uses of record linking as applied simply to the familiar kinds of *ad hoc* studies of limited size and duration. The question arises whether it is realistic to go beyond this and to consider using, for scientific purposes, all of the pedigree information

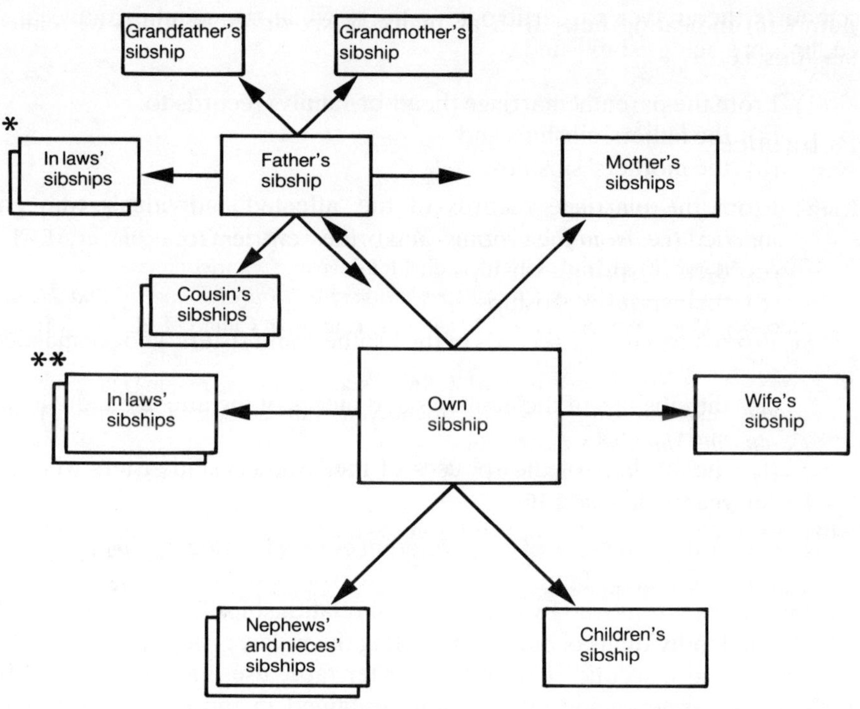

Fig. 1.5 Scanning the master file for related sibships. * Those of the paternal uncles and aunts by marriage; ** those of brothers' wives and sisters' husbands.

gathered routinely for whole populations through the vital registration systems, of doing so on a continuing basis, and of adding an increasing amount of medical documentation as time goes on.

Clearly, the cost would appear large if it were paid wholly from budgets for scientific research. But this would not necessarily be the case, because the information that is unlocked by linking and integrating the files into individual and family histories has many statistical and administrative uses, as well as other scientific uses beyond those of the geneticist.

Those geneticists who attempt to apply the methods of record linking will be in a particularly good position to see a variety of possible uses for the linked files and to develop procedures that will serve more than one purpose. Their own long-term interest may be furthered most where they exploit the fact that there are other potential users.

Of course, with time the various files of routine records will, to an increasing extent, be linked and integrated anyway for administrative purposes, whether or not scientists take an interest in the matter. But the only way to ensure that scientific by-products will come out of this trend is for the scientists themselves to participate actively while the administrative procedures are being established.

References

KENNEDY, J. M., NEWCOMBE, H. B., OKAZAKI, E. A., AND SMITH, M. E. (1964). *List Processing Methods for Organizing Files of Linked Records*. Document AECL–2078. Atomic Energy of Canada Ltd., Chalk River, Ontario.

——, ——, ——, —— (1965). *Computer Methods for Family Linkage of Vital Health Records*. Document AECL–2222. Atomic Energy of Canada Ltd., Chalk River, Ontario.

NEWCOMBE, H. B. AND KENNEDY, J. M. (1962). Record linkage: making maximum use of the discriminating power of identifying information. *Commun. Assoc. Computing Machinery* **5**, 563–6.

——, ——, Axford, S. J., and James, A. P. (1959). Automatic linkage of vital and health records. *Science* **130**, 954–9.

SMITH, M. E., SCHWARTZ, R. R., AND NEWCOMBE, H. B. (1965). *Computer Methods for Extracting Sibship Data from Family Groupings of Records*. Document AECL–2530. Atomic Energy of Canada Ltd., Chalk River, Ontario.

2

Methods and technology of record linkage: some practical considerations

L. E. GILL AND THE LATE J. A. BALDWIN

In Chapter 1 the original theory and techniques of record linking to create files of person and family records were described. The efficiency of record linking, using the large capacity and high speed of modern computing technology, is now such that systematic linking of processable records on a large scale is feasible as a routine. In this chapter we shall discuss some of the practical aspects of record linking, drawing in particular on our experience with the Oxford Record Linkage Study (ORLS).

Matching and linking: practical considerations

Once potentially linkable records have been found (i.e. in the searching operation), the linking process involves two stages: matching records and linking them together. *Matching* is the process of comparing pairs of records to determine whether they should be linked. The members of the pairs may contain data about the same episode of illness (for example, an admission and discharge record), or about different episodes relating to the same individual, or they may be about different individuals related in specified ways. For simplicity, we shall focus on matching records for different episodes which relate to the same person. *Linking* is the process by which pairs of correctly matched records are brought together in such a way that they may be treated as a single record for one individual.

The fundamental requirement for correct matching is that there should be a means of uniquely identifying the person (or episode or family) on every document to be linked. In some records, such as the parts of a hospital admission and discharge abstract, unique identification of an event can be virtually guaranteed by means of a single item, such as a form number. In these cases, matching is a relatively simple process. Difficulties arise, however, where the records do not include unique identifiers. In these instances, matching depends on achieving the closest equivalent to unique identification by using several matching variables, each of which is only a partial identifier, but which in combination provide a match which is sufficiently accurate for the intended uses of the linked data. The difference between matching by means of a single unique identifier and matching by means of combining several identifiers is of such practical importance that separate consideration is necessary. The former can be styled as 'all-or-none matching' since the pairs of records compared either do or do not match; in other words, there is no question of partial agreement. The use of several identifiers to match records is called 'probability matching' since the object is to estimate the probability that the records in each pair refer to the same individual.

All-or-none matching

The principal requirement for all-or-none matching is the availability of a characteristic of the person that is fixed, easily recorded, and unique but also readily accessible and verifiable. Few, if any, identifiers meet all these specifications, though several come sufficiently close to be usable. The perfect identifier would be an integral feature of the individual, such as a personal trait or a group of traits which together form a unique set. Alternatively, unique characteristics may be assigned to individuals at birth or by means of a highly reliable matching procedure.

Personal traits

The fingerprint is probably the most obvious of these. Its almost universal use in police records attests to its efficacy as an identification device. However, it is precisely for this reason that the fingerprint is inappropriate for health records. An alternative personal trait is the acoustic wave pattern of the voice, or 'voice-print'. Although this is probably different for every individual and in theory is easy to record, the difficulties of making standard recordings, the cost, and the technical requirements for interpretation mean that it is unlikely to be used for medical record linkage. Similarly, although typing of tissues or 'gene-prints' may in future become standard practice in medical care, it seems unlikely that these techniques would be practicable for identification purposes. In summary, personal traits are not readily applicable as identifiers in medical record linkage.

Assigned characteristics

A close approximation to an ideal all-or-none matching device can be achieved by using a series of numbers large enough to encompass all the members of the population. Many numbering systems are in use and these may be classified into three broad groups:

(1) *Serial numbering systems*—in which a unique number is assigned to each individual from a central allocating point. These systems can be very reliable if the number incorporates a checking device that permits verification. Serial

numbering systems have three main advantages: they are simple to use; easy to automate; and do not depend on non-unique features of the individual. Conversely, strict control is needed over the allocation of numbers and the number cannot be reproduced on the basis of an individual's characteristics.

Serial numbering is often used for patients' records in unit medical record systems (Baldwin and Gill 1982). Here allocation can be strictly but simply controlled by the use of a central number 'bank', which may be a book or a computer file, showing the allocation of each number and the next one available in the sequence. Records may be filed according to the terminal digit rather than in simple sequence and this can be used to minimize the facilities needed for searching, retrieval, and filing (Sargeant 1977). Automatic numbering machines may be used to produce labels, to identify forms and other documents, and to locate the master index card for each patient. These cards can in turn be employed to retrieve a patient's number from a file ordered on the basis of name. Difficulties often arise, however, when a unit medical record system becomes very large and, consequently, the numbers allocated are also large. It is known that numbers with more than three or four digits are prone to transcription errors and should be verified by means of a check character. Moreover, as the system expands, errors become progressively more difficult to detect and correct.

(2) *Derived numbering systems*—these depend on universal and readily available non-unique characteristics of the individual from which the number is generated. The most important advantage of these systems is reproducibility since numbers can be compiled at any site and at any time without reference to a central allocation point. The disadvantages are relatively low reliability, since they depend on the accuracy of the individual's stated characteristics, and the risk that two individuals will actually or reportedly have the same set of non-unique characteristics and thus will be assigned the same number.

A well-known example of a derived numbering system is the Midland Personnel Number (Hogben and Cross 1948). This is an 11-digit number compiled from codes for the individual's surname (or the maiden name of a married woman), first forename, date of birth, and birth rank. Several variations of the original 'Hogben' number were tried and it is still recommended for small files of records, such as those in general practice (Crombie 1981). Research by the ORLS, however, indicated that the reliability of reporting the data items was low and that the Hogben number differed in about 15 per cent of all pairs of records for the same person (Acheson 1967). Moreover, the discriminating power of the number was not adequate since the birth rank item was difficult to collect and the surname code reduced rare names to the same weight as common names.

(3) *Composite numbering systems*—combinations of the serial and derived systems that use a central allocating point to obtain part(s) of the number and

non-unique characteristics to derive the other part(s). Examples include driving licence number and identification number for members of the armed forces. Although it is often difficult to devise satisfactory checking techniques for these composite systems, they do have the advantage of a reduced risk of any one number being given to more than one individual.

Errors

All three types of system described above are prone to errors in the recording of numbers, whether recording is by speech, handwriting, or keying. There are four basic groups of errors. Firstly, *transcription or substitution errors* are those in which one or more correct digits are replaced by incorrect characters through mishearing, misreading, or miskeying. This type of error is by far the most common, and in the experience of the ORLS, accounts for about 86 per cent of errors in the recording of numbers but applies to less than one per cent of all records. A common cause is poor definition of handwritten and printed digits, such as confusion between 1 and 7 or 3 and 8. Secondly, there are *transposition errors* in which two correct digits are transposed in adjacent positions (single transposition) or across an intervening correct digit (double transposition). In the ORLS, the former contribute approximately eight per cent of all errors and the latter one per cent. Thirdly, *shift errors* occur when a whole number is moved one digit position or more to the left or right by addition or omission of consecutive zeros. These tend to account for less than one per cent of errors in most number systems. Finally, there are several other types of error which fit none of these categories and which together constitute less than five per cent of all data preparation errors in the ORLS.

Check characters

One means of reducing all types of error in the recording of numbers is to incorporate a checking device. The most widely used of these is based on the inclusion of a prime number or modulus and involves assigning a number or weight to each digit position of the record number to be checked. Each digit value is multiplied by the weight for its position and all the products are summed. The sum is then divided by the modulus and the remainder of the division is used as the check character. Applying this calculation to a reported record number gives a check character value which can then be compared with the known check character. If the two values are not identical, an error has occurred either in the number or in the reported check character.

Table 2.1 indicates that the larger the value of the modulus, the higher the proportion of all possible errors detected, and therefore the selection of the modulus is clearly critical. The number of possible check characters is one less than the value of the modulus and if there were no practical limits on the number of check characters, extremely high error-detection rates could be attained. However, in practice it is often preferable to use only one check

Table 2.1 Error detection rates (per cent) in self-checking serial
numbers using remainder method of check character derivation

Modulus	Error Type*				
	Transcription/ Substitution	Transposition		Shift and Other	Total
		Single	Double		
7	93.3	93.3	73.3	85.7	92.72
11	100	100	100	90.9	99.55
13	100	100	100	92.3	99.62
23	100	100	100	95.6	99.78
97	100	100	100	99.0	99.95

* See text for definitions.
Source: Post Office (1962).

character. If an alpha character is adopted there is a maximum of 22 values to choose from using modulus 23, which is the highest prime number available. Alpha check characters also have the advantage of being readily distinguishable from the numerals in the record number.

The use of check characters to reduce errors in the recording of numbers may introduce further errors, since there is a risk of transcription mistakes. Although errors may be less common in transcribing a single alpha character than in transcribing one or two numeric digits, certain letters are prone to misinterpretation and are best avoided, particularly if they are handwritten. In one experiment (Kenrick and Jefferson unpublished) involving the hand transcription of 113 000 numbers using a single alpha check character based on modulus 23, 1 614 (1.4 per cent) numbers were found to be in error. Of these errors, 1 130 (70 per cent) were in the check character alone, 371 (23 per cent) were in the number alone, 19 (1 per cent) were in both the number and the check character, and the balance of 94 (6 per cent) were keying errors caused by defective handwriting of the transcribed numbers. Analysis of the errors for each alpha character used gave values from 16 per cent for letter U down to 2 per cent for letter T. Only 14 letters had error rates less than 4 per cent. This research provides strong support for the use of both modulus 13 and the following 12 alpha check characters: A, B, C, D, E, F, K, L, R, S, T, and X.

It will be apparent from the above discussion that, whatever modulus is used, the allocation of check characters will follow a cyclic pattern dependent on the weights selected. Detection of some types of error is affected by the arrangement of weights and, ideally, both arithmetic and geometric progressions should be avoided. A technique for allocating weights is available that approaches this ideal, whereby no three adjacent values form an arithmetic or geometric progression and no two adjacent digits sum to the value of the modulus (Beckley 1967). For a general discussion of the theory of check digits, the reader is referred to Post Office (1962) and Wild (1968).

The calculations involved in the use of check characters clearly require machine assistance. Special machines have been marketed to carry out the procedure, usually based on modulus 11, but many hand-held programmable calculators now provide a cheap and flexible alternative. Nevertheless, in most applications, the check calculation is carried out automatically at the time of entry of record numbers into the computer system. Handwritten transcription of numbers may be avoided by the use of pre-printed labels produced by the computer and by the pre-printing of numbered form series. However, unless some means of optical reading is used, such as bar-coding, numbers will have to be keyed manually for computer entry and here verification is essential.

The UK National Health Service number

In the United Kingdom, the only assigned number that might be used on a large scale for all-or-none matching of medical records is the National Health Service (NHS) number. This provides a unique, fixed identification number for almost the whole population. The number originated from national registration in September 1939 and was adopted by the NHS in 1948. Since then all persons born in the country and all persons entering the country and registering with a general practitioner have been given an NHS number. This forms the basis of the National Health Service Central Register for England and Wales, and similar but separate registers are maintained for Scotland and Northern Ireland. The primary purpose of these registers is to help the Family Practitioner Committees maintain lists of general practitioners' patients. The central registers are updated by the addition of entries for births (from copy birth certificates) and immigrants who register with family doctors, and by the removal of deaths (from copy death certificates) and emigrants. Changes of family doctor are also indicated. In performing these functions, the central registers use the NHS number for matching whenever possible.

As a record-matching device the NHS number has two major drawbacks. Firstly, it is not generally available on medical documents. Although the number is found on the majority of general practitioners' records and on all birth certificates, this is true of only a proportion of death certificates and of an even smaller proportion of hospital records. The hospital service has not made effective use of the NHS number for its own record-keeping largely because there is no specific requirement for patients to use or know their number. In the Oxford Region, for example, although the NHS number is routinely requested from patients being admitted for in-patient care, the number is actually obtained for only about one-quarter of all admissions. The second important problem with the NHS number for matching is its complex and variable format. The number can vary in length up to a maximum of 11 characters in England and Wales and 15 in Scotland, and consists of variable numbers of alpha and numeric characters separated by punctuation. Moreover, the format of the NHS number has been changed repeatedly (Farmer and Cross 1973). For matching purposes, therefore, reliance cannot be placed on NHS numbers reported from memory; ideally the number should always be taken from a medical card or other official document, such as the general practitioner's record. Transcription of these long and complex series of arbitrary characters is obviously prone to error, confounded by the fact that in Scottish numbers the position of punctuation slashes and full stops is critical. Although Farmer and Cross (1973) have specified an algorithm which is effective in detecting numbers beyond the valid range, a more precise checking device is not available.

Despite these problems, the NHS number is valuable for record linkage purposes largely because of its all-or-none matching quality. If there is exact correspondence between NHS numbers on two different documents, a match may be presumed with little or no further checking. However, transcription errors will sometimes cause missed matches. Less likely are false positive matches arising from exact correspondence between an incorrect and a correct number or two incorrect numbers. When records do contain the NHS number, computer matching can be rapid, reliable, and inexpensive. To summarize, there is little doubt that the NHS number could be collected and used on a far wider scale than at present if record linking became routine in medical information systems in the United Kingdom health service.

Probability matching

Whatever method of matching is used, all possible pairs of records should be compared. It will be apparent that, even with quite small files of records, the number of comparisons could become very large and impracticable in terms of complexity, time, and cost. The initial procedure in all matching techniques is therefore to arrange the records in such a way that comparisons are made in the most efficient way possible.

In all-or-none matching, the file should be ordered sequentially on the values of the single identifier. This achieves the maximum efficiency in comparing records. In probability matching, however, there is no single unique identifier on all records. It is, therefore, necessary to group together the records that are likely to refer to the same person, thereby reducing the time spent searching the file, and to make comparisons only within these groups. Clearly the reliability and efficiency of the matching procedure is highly dependent on the way in which the initial grouping of 'file-blocking' step is carried out. One important consideration is that the number of records in each block is small enough to avoid too many unproductive comparisons and yet large enough to prevent records for the same individual falling into different blocks and so failing to be compared. The balance between the number and size of blocks is particularly important when large files are being matched. The selection of variables to be used for file-blocking is, therefore, critical and will be discussed before considering the comparison and decision-making stages of probability matching.

File-blocking items

Any variable that is present on all the records to be matched could be used to divide the file and thus to reduce the number of comparisons. The item 'sex', with two values (or three if not known is allowed), would roughly halve the number of comparisons, whilst the item 'year of birth' would create a larger number of groups, often over 100, and the combination of both these variables would give still better discrimination. Nevertheless, if there is a risk that the items chosen are wrongly recorded, causing the records to be assigned to the wrong file-block, then potential matches will be missed. Similarly, items that are likely to change their value from one record to the next for the same person (such as home address) are also unsuitable for file-blocking. Thus, the items used for file-blocking should be universally available, reliably recorded, and permanent.

There are three numerical measures often used to evaluate items used for file-blocking, namely the coefficient of specificity, discriminating power, and merit ratio. These were discussed in detail in Chapter 1. Table 2.2 gives examples of items which may be employed for file-blocking and their relative merits. In practice, it is almost always necessary to use surnames, combined with one or two other ubiquitous items, such as sex and year of birth. For this reason, considerable attention has been given to coding surnames in ways which reduce or eliminate the effects of variations in spelling and reporting, and which 'compress' names into fixed length codes.

Name compression

Early work on name compression was concerned with mechanisms for deriving unique codes for individuals from names and other variables suitable for all-or-none matching; this has already been described. By contrast, name compression codes are *now* used mainly to group together variants of surnames, for the purposes of blocking and searching, in order that effective match comparisons can be made using both the full name and other identifying data despite misspelt or misreported names.

The main requirements of a name compression code are that all variations of a name are included in the same group, but dissimilar names are excluded. Moreover, the groups should be of a limited and fairly uniform size. A large number of simple codes have been used, often based on a predetermined selection of letters from the name, such as the first, third, fifth, and final characters. Whilst this is moderately effective in overcoming some common misspellings, they tend to place an unsatisfactory number of dissimilar names in the same group and are not suitable for use with large files of records. A major advance in name compression was made by applying the principles of phonetics to group together classes of similar sounding groups of letters and

Table 2.2 Relative merits of the characteristics of identifying items for person matching

Item[1]	Characteristics				Notes
	D^2	S^3	E^4	A^5	
Present surname	Av[6]	Av	Hi[7]	Hi	Always available.
Birth surname	Av	Hi	Hi	Av	When used with mother's birth surname it is highly discriminating.
Forenames	Av	Hi	Hi	Hi	Use at least two forenames. If more than two, use all if possible.
Date of birth	Av	Hi	Hi	Hi	Should be checked with age at interview and recorded as date, month, year.
Place of birth	Av	Hi	Av	Av	Literal place of birth is often given as maternity hospital. This has less discriminating power.
Sex	Lo[8]	Hi	Lo	Hi	The only item rarely presenting ambiguity.
Civil state	Av	Lo	Av	Hi	Can be expanded with advantage, i.e. married once, more than once, etc.
Address	Av	Lo	Av	Hi	Full postal address, plus post office postal code. Usual address should be recorded in addition to present address.
NHS number	Hi	Hi	Hi	Lo	Must be copied from an official document, such as medical card. This should not be confused with the national security number.
Family doctor	Av	Lo	Av	Hi	Commonly required, useful for identification. Record as surname, initials, and address.
Date of marriage	Av	Hi	Av	Lo	Valuable for family record linkage.
Mother's birth surname	Av	Hi	Av	Av	Highly discriminating when used with birth surname. Valuable for family record linkage.
Spouse's birth surname	Av	Hi	Av	Lo	Valuable for family record linkage.
Period of residence	Av	Lo	Av	Lo	Valuable check item with address, especially in very mobile populations.

[1] All items, except civil state, date of marriage, mother's birth surname, and period of residence are used by the ORLS.

[2] D = Discriminating power.

[3] S = Stability, i.e. liability to change. Where stability is low or average, provision must be made for recording changes and date of change.

[4] E = Liability to error.

[5] A = Availability; likely in situations where attempts are made to collect the item.

[6] Av = Average.

[7] Hi = High.

[8] Lo = Low.

Source: Baldwin (1972).

thus similar sounding names. The best known of these codes was devised in the 1920s by Odell and Russell (Knuth 1973) and is known as the Soundex Code; this was discussed in Chapter 1. Other name compression algorithms are considered by Dolby (1970).

The Soundex Code has been widely used in medical record systems despite its disadvantages. In particular, it fails to bring together some common variants of names, such as Thomson and Thompson, and it is also unsatisfactory for certain non-English names. Moreover, many of the groups formed by Soundex contain too high a proportion of dissimilar names and some groups are too large for maximum efficiency. Equally, the assumption that the first letter of a name is reliable is not always correct. Several modifications of the original Soundex algorithm have been proposed to overcome these difficulties. One that is particularly useful for Scottish names is the SINGS code (Soundex, Initial Numericise Granick, Smith) (Smith 1968).

A further advance in the development of modern name compression techniques took place with the transformation of consonants within names to specific combinations of *both* vowels and consonants (see Dolby 1970). Among several algorithms of this type, that devised by the New York State Information and Intelligence System (NYSIIS) has been particularly successful, and has been used in a modified form by Statistics Canada for an extensive series of record linkage studies (Lynch and Arends 1977). A more recent development by the ORLS (Gill 1981), referred to as the Oxford Name Compression Algorithm (ONCA), uses the modified NYSIIS method of compression as an initial or preprocessing technique. The transformed and partially compressed name is then Soundexed in the usual way. This two-stage technique has been used successfully for blocking the files of the ORLS, and it overcomes most of the unsatisfactory features of pure Soundexing whilst retaining a convenient four-character fixed-length format.

The comparison and decision-making steps

Methods for linking two or more records belonging to the same person depend essentially upon comparisons between the various items of identifying information. Based on the discriminating power and reliability of each of these items, numerical values can be calculated for the amount of agreement or disagreement. The values indicate the likelihood that a pair of records refer to the same individual, and are expressed as binit weights. These weights simulate the subjective judgment of a clerk and provide a means by which the assurance of a match is increased; the calculations involved were considered in detail in Chapter 1. The procedure for deciding if two records belong to the same person was first developed by Newcombe, Kennedy, Axford, and James (1959). The decision is based on the total binit weight, which is derived by summing algebraically the individual binit weights calculated from the comparisons of each identifying item available on the new input-data file and on the master file of records that have already been matched. The total weight represents an index of the probability that two records match. Table 2.3 shows the availability of identifying items on the ORLS 16-year (1963–1978) file and Table 2.4 indicates the binit weights for some of these items.

Typically, two distinct distributions are created when the number of

Table 2.3 Percentage availability of identifying items on discharge and death records on the ORLS 16-year (1963–1978) file

Identifying Data Items	General Hospital Discharges		Death Records	
	1963–1970 (n = 375 079)	1971–1978 (n = 1 149 295)	1963–1970 (n = 47 123)	1971–1978 (n = 119 260)
Present surname	100.00	99.9	100.0	99.9
Birth surname	32.8[1]	81.5	81.0	92.4
First forename	99.9	99.9	99.9	99.9
Second forename	57.4	32.6	70.7	73.9
NHS number	26.4	16.0	16.7	9.7
Place of birth (code)	78.6	71.3	39.4[2]	94.9
Home address (literals)	99.9	96.2	98.7	97.8
Home address (code)	99.9	96.2	98.7	97.8
General practitioner	98.3	95.4	42.0[3]	90.0[4]
Sex	100.0	99.9	99.9	99.9
Date of birth	99.9	99.7	100.0	100.0

[1] Only collected for married women 1963–1970. 32.8 is, therefore, approximately the percentage of married women on the file for that period.

[2] Percentage of death records where place of birth could be ascertained from a previous discharge record.

[3] Not recorded for hospital deaths.

[4] Includes domiciliary and hospital deaths.

record pairs being matched is plotted against the total calculated binit weights. One distribution is for pairs that are matched and the other is for pairs that do not match (Butler 1971) as shown in Fig. 2.1 for the ORLS 16-year file. The area of overlap between the two distributions represents an area of *possible* matches. Pairs of records with weights lying within the area of overlap fall into one of four possible categories:

(1) The records are matched and belong to the same person.
(2) The records do not match and do not belong to the same person.
(3) The records are matched but do not belong to the same person, i.e. false positive matches.
(4) The records are not matched but do belong to the same person, i.e. missed matches.

In principle, one could scrutinize all pairs of records that fall within the area of overlap. In practice, this would result in an inordinate amount of clerical checking. Alternatively, one could take a cut-off—at, say, the point of intersection of the two distributions (binit weight 11 in Fig. 2.1)—above which all matches would be accepted as records belonging to the same person and below which all 'non-matches' would be deemed to be records

Table 2.4 Match weights used by the ORLS for the 16-year (1963–1978) file

Identifying Data Item	Score in binits[1]		
	Exact Match	Partial Match	No Match
Surnames: birth	+2 to +6	−3 to −6	−6
present[2]	+2 to +6	−3 to −6	−6
mother's birth	+2 to +6	−3 to −6	−6
Forenames[3]	+2 to +7	equivalent +2 initials +1	−5
NHS number	+7	*[4]	−2
Place of birth (code)	+4	+2	−4
Home address[5]	alpha +7 numeric +5	*	−3
General practitioner (code)	+4	+2	−2
Sex[6]	+1	*	−10
Date of birth: year	+5	−1 to −5	−5
month	+4	*	−4
day	+5	*	−4
Hospital and hospital unit number	+7	*	−2

[1] Items, other than those below, recorded as not known, left blank, or in error are scored 0.

[2] If present surname is not known, the record is appended to the file but cannot be matched with any other record.

[3] Forename entries, such as boy, girl, baby, infant, twin, or not known are scored −10.

[4] * = not permissible or not possible.

[5] No fixed abode is scored 0.

[6] Sex not known, blank or in error is scored −10.

relating to different people. In the ORLS we normally set a range of values of binit weights within which pairs of records will be subject to clerical scrutiny. This range is delimited by the upper and the lower preset thresholds shown in Fig. 2.1. The values selected for these thresholds are, of course, arbitrary but are chosen as a result of considering the following different objectives.

(1) The minimization of false positives, at the risk of increased missed matches.

(2) The minimization of missed matches, at the risk of increased false positives.

(3) The minimization of the sum of false positives *and* missed matches.

The relative importance of these objectives, and therefore also the cut-off point and preset thresholds, will vary between studies according to the intended uses of the linked data.

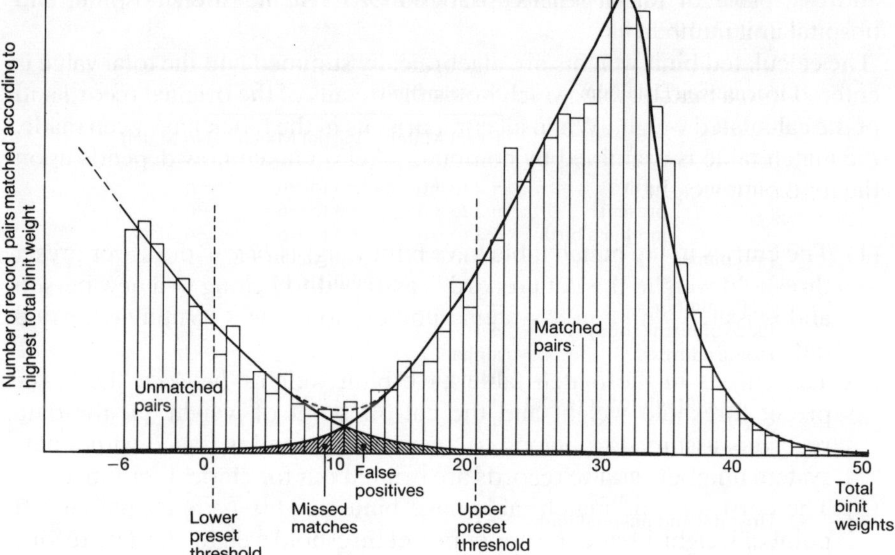

Fig. 2.1 Frequency distribution of total binit weights for record pairs (matched according to total binit weight) on the ORLS 16-year file.

Matching and linking in the Oxford Record Linkage Study

In the introduction to the book, the origin and development of the ORLS was described. It was indicated that the data files contain abstracts of clinical records for general hospital in-patients, maternity in-patients, psychiatry in-patients and out-patients, and for vital events (birth and death abstracts). Prior to matching, each type of record that enters the files is re-formated to a common layout, edited and corrected, and individually 'protected' (by enciphering) for data security. A match header is added to each record, in which the ONCA compression of the birth surname and present surname is stored, together with sex and year of birth.

Matching

To match incoming records to an existing linked file an extract file, called the master match file (MMF), is taken from the already linked data file. This extract is blocked on the basis of sex, a compressed birth surname, and year of birth order and the new input data file is treated in a similar way. Each new data record is compared against each of the existing records in the corresponding MMF block. Binit weights are calculated for the comparison between birth surname, present surname, forenames, sex, date of birth, street address, place of birth, general practitioner, NHS number, hospital, and hospital unit number.

The calculated binit weights are algebraically summed and the total value is entered into a match table, which holds the details of the original record and of the calculated weight. When all comparisons in the block have been made, the match table is examined by computer. Action taken now depends upon the total binit weight.

(1) The entries in the match table have binit weights *below* the lower preset threshold weight: the data record is assumed to belong to a new person and is issued with a new system number; no clerical scrutiny is carried out.

(2) The entries in the match table have binit weights *between* the lower preset threshold weight and the cut-off point of weight 11: the data record is assumed to belong to a new person and is issued with a new system number; *all* the records are printed out for clerical scrutiny.

(3) The entries in the match table have binit weights *between* the cut-off point of weight 11 and the upper preset threshold weight: the data record is given the system number of the record in the MMF block to which it matched with the highest weight; all the records are printed out for clerical scrutiny.

(4) The entries in the match table have binit weights *above* the upper preset threshold weight: the data record is given the system number of the record to which it matched with the highest weight; a record is only printed out if it matched equally with two or more records with different system numbers.

If the data record is 'judged' to be a new person, an extract of the record is added to the appropriate MMF block. In addition, if more than one record for a particular new person is being added to the file in the same computer run, then each record should receive same system number, thereby ensuring that they are matched together. A copy of this matched record is appended to the MMF to ensure that further potential records in the run for the person are given the same system number.

At the end of a match run, the data file is split into two separate files. One contains records which can be accepted as matched plus those which can be accepted as new individuals, i.e. without an existing record with which to match. The other file contains records which are doubtful matches and require clerical checking. The clerk's special knowledge of the identifying data and local circumstances help in checking and, if necessary, correcting the computer decision. This procedure can reduce the number of missed matches and false positives to tolerably low levels.

The accepted matches are sorted on the basis of system number and date of event, and are merged with the master file. Each new batch of data to be submitted to the linked file requires two match and linking computer runs.

The records are blocked firstly on the basis of birth surname and secondly on present surname. Periodically the MMF is matched internally by blocking on completely different variables, such as forename. This process may further reduce the number of missed matches by about one to two per cent by bringing together records which, owing to errors in the variables originally used for file-blocking, were initially entered in different MMF blocks.

Linking

Once a data record has matched with a record already on the MMF and has been given the same system number, logical checks are performed to identify some of the false positive matches. These checks may produce a printed output for clerical scrutiny. All the records for a person are then assembled in temporal order, ending with a death record when relevant. This sequence of records is examined for overlapping periods of care or 'changes' of sex, and for events 'occurring' after a death record or before a birth record. Any of these inconsistencies in a set of records results in clerical checking and corrections are made where possible.

Conclusion

The procedures for linking records described in this chapter, though clearly open to further development, are suitable for widespread use. These procedures can be carried out using medium-sized computers, given sufficient memory and backing store, and are reliable and efficient. Indeed, the cost of linking is only a small fraction of the cost of the original data collection and file compilation, and is modest in relation to the increased value of the material that can be extracted. Record linkage is simplified and the costs much lower when the records are compiled with the specific object of linking. In research, however, it is more usual to attempt to link together records produced primarily for other purposes—notably the recording of an event, such as hospitalization, for immediate operational purposes—and as a consequence, the material available to effect linkage is often less than ideal. This chapter has considered some of the methods of record linkage which take account of these problems.

References

ACHESON, E. D. (1967). *Medical Record Linkage.* Nuffield Provincial Hospitals. Oxford University Press.
BALDWIN, J. A. (1972). Linked record health data systems. *The Statistician* **21**, 325.
—— and Gill, L. E. (1982). The district number—a comparative test of some record-matching methods. *Community Medicine* **4**, 265.

Beckley, D. F. (1967). An optimum system with 'Modulus 11'. *Computer Bulletin* December, 213.

Butler, A. R. (1971). Record linkage: adaptive decision techniques in file processing. In *Information Processing* (North Holland Publishing Co., Amsterdam), 1466–71.

Crombie, D. L. (1981). Record linkage in automated systems. *Journal of the Royal College of General Practitioners* **6**, 325.

Dolby,J. L. (1970). An algorithm for variable-length proper-name compression. *Journal of Library Automation* **3/4**, 257.

Farmer, R. D. T. and Cross, K. W. (1973). The National Health Service number. *British Journal of Preventive and Social Medicine* **27**, 257.

Hogben, L. and Cross, K. W. (1948). The statistical specificity of a code personnel cypher sequence. *British Journal of Social Medicine* **2**, 149.

Gill, L. E. (1981). Revised record linking method. (Unpublished paper) Oxford Record Linkage Study.

Kenrick and Jefferson Ltd. (undated). The Language of Numbers—Check Digit Numbering. (Unpublished paper) England.

Knuth, D. E. (1973). Sorting and searching. In *The Art of Computer Programming* Vol. 3, p. 391. Addison-Wesley Publishing Co. Inc., United States.

Lynch, B. T. and Arends, W. L. (1977). *Selection of a surname coding procedure for the Statistical Reporting Service Record Linkage System.* United States Department of Agriculture, Washington D.C.

Newcombe, H. B., Kennedy, J. M., Axford, S. J., and James, A. P. (1959). Automatic linkage of vital records. *Science* **130**, 954.

Post Office (1962). *The function and mathematics of check digits.* Report 170 of the Central Organization and Methods Branch of the Clerical and Buildings Department, London.

Sargeant, F. W. N. (1977). Filing systems and indices. In *Medical Records* (ed. B. Benjamin). William Heinemann Medical Books Ltd., London.

Smith, A. (1968). Preservation of confidence at the central level. In *Record Linkage in Medicine* (ed. E. D. Acheson). Proceedings of International Symposium, Oxford, July 1967. E. and S. Livingstone Ltd., London.

Wild, W. G. (1968). The theory of modulus N check digit systems. *Computer Bulletin* December, 309.

3

Analytical methods for time-sequenced linked records

J. GOLDING, S. P. VIVIAN, C. J. BAINES,

AND THE LATE J. A. BALDWIN

Introduction

Extensive collections of linked individual and family medical records relating to defined populations at risk call for appropriate methods of statistical analysis. The data files we shall consider take the unit of record as the individual or family, with information on a variable number of different types of event or episode ordered in time sequence. In this chapter we shall describe some of the analytical methods suitable for the main applications of linked data, including measures of incidence and recurrence in individuals, quantitative estimation of outcome, measurement of statistical association between diseases, case-control studies, and population genetics and other analyses appropriate to the study of families. Firstly, it is important to consider the nature of linked observational data and the precautions necessary for its use.

The nature of data to be linked

Geographic and social variation

Although some data items in linked records can be dealt with easily, such as calculation of age from date of birth or the date at which a particular event occurred, there are several key items which may change for a given individual in an unpredictable manner and so present the statistician with difficult decisions. Such items include area of residence, marital status, occupation, and social class.

At each successive event or episode, the linked records system may acquire fresh information about the demographic and social features of each individual. All these should be stored, together with the dates on which the information was recorded. Changes can then be ascertained and related to the particular time period of an analysis. The simplest approach is to select those demographic and social variables that were valid at a specific date relating to the key episode of the analysis. The date may be that on which a particular diagnosis or operation was recorded. However, each problem or analysis should be considered *de novo* to determine the most logical approach.

Clinical data

Clinical information, notably diagnoses and surgical operations, cause additional complications. Not only may these data items change from one episode to another, but there may also be a number of diagnoses or operations in each record, and their order of entry may vary. The International Classification of Diseases, Injuries, and Causes of Death (ICD) has specified the rules for entry of diagnoses on mortality records for many years, but the Ninth Revision (World Health Organization 1977) is the first to include rules for the selection and ordering of diagnoses in statistical abstracts of morbidity records. Until these came into general use, the ordering of diagnostic entries followed no strict rules. General recommendations, such as entering first the condition causing admission, or the most serious diagnosis, or that for which the patient was primarily treated, are rarely implemented reliably. The first entered condition may just be the first encountered by the abstractor.

Furthermore, few statistical abstracts of morbidity records distinguish between a newly diagnosed condition and a recurrence of an earlier disease. Mental hospital records are exceptional in this respect, the high frequency of readmission having led to early use of the distinction between first admission with a particular disorder and readmission. Even here it is not always clear whether first admission means to the reference hospital, in the patient's life, or even just within a specified time period.

Further complications may arise due to changes in the classification of data items which inevitably occur when the interval between episodes is protracted. An obvious example of this is provided by the ICD, which has been revised repeatedly to accommodate changes in the logic and discriminatory power of nosology, with consequent alterations in the meaning of codes.

Analyses of linked files

Incidence, prevalence, and recurrence

Perhaps the simplest statistics to be produced from a file of linked records are incidence or period prevalence rates. These must be defined clearly, with meaningful numerator and logically valid denominator. Any comparison of rates must always compare like with like. Numerators are usually obtained by counting the number of individuals in the population at risk to whom one or more defined events occur over a specified period of time. The denominator should consist of the number of persons in the population at risk during the same time period. The most meaningful interpretation of rates will occur if the population at risk consists of all persons resident in a defined geographic area. An alternative would be the population of persons on the linked file.

Rates will usually be given in terms of specified characteristics, such as sex, age, and diagnosis. Age changes with time and ordinarily will mean the age in completed years at the time of the episode. A range of options is likely to be available for counts of individuals by diagnosis. If the purpose is to obtain, say, *annual discharge rates* by diagnosis (often loosely referred to as annual incidence), the correct course is likely to be to count each individual with a specified diagnosis once for each of the years for which he was recorded as having such a diagnosis. Thus if he had 8 episodes with diagnosis D, 1 in year 1, 3 in year 2, 1 in year 3, and 3 in year 8 he would contribute 1, 1, 1, 0, 0, 0, 0, 1 respectively to the numerator of the rates in each year. Alternatively, for a *point prevalence rate*, the numerator would be those individuals in care with a given diagnosis on some specified date. To obtain the *period* or *administrative prevalence rate*, one would add to the numerator obtained for the point prevalence rate those individuals who entered care with the same condition during a subsequent defined period, such as a year.

In any study of incidence or prevalence rates, it must be decided in advance whether the analysis should be restricted to the 'primary' diagnosis position. Clearly, the numerator and thus the rates obtained are likely to differ widely according to whether it is the primary diagnosis, or any diagnosis that is used. It is absolutely essential that both user and analyst agree precisely on what is most appropriate.

Rates of recurrence entail further complication. Given an individual with diagnosis D the problem is to compute the risk of having a further episode with diagnosis D. Alternatively, the analyst may find the mean number of admissions with D within a given time period to be more meaningful. Sophisticated concepts, such as that of the cohort life table, are not widely understood by clinicians, administrators, and others and a more easily digested formulation is likely to be required, whatever its imperfections.

Though of limited value, an overall *readmission rate* may be required for study of patient movement through hospitals. The best representation of this may be a count of all episodes, the denominator being the number of persons concerned. Alternatively, it is quite useful to give the frequency distribution of number of persons by number of admissions. Neither of these tabulations show the extent to which recurrence or readmission for a given condition occurs. Counts of individuals having more than one episodic record carrying the same diagnosis would be appropriate to this purpose.

The main disadvantage in all these simple counts of recurrence is that the periods over which individuals are at risk are not held constant, except in the general sense that if occurrences are evenly distributed by time and if there are no fundamental changes in the population at risk, the mean period at risk will be half the overall time period. For administrative applications in health services this deficiency may be of less importance than comparability between calendar periods.

For clinical applications, such as comparisons of performance, where comparability of period at risk is more desirable, a simple form of cohort analysis can be considered. Here individuals may be followed for some fixed period, such as one year, after an initial episode carrying a specified diagnosis, and counts obtained of individuals for whom recurrence is recorded. Successive time periods may be compared with corrections for varying periods at risk caused by death, migration, and differing file lengths. At this degree of complexity, it is preferable to adopt a full cohort life table approach (see below).

Mortality rates

While mortality rates by underlying cause of death can be extracted as easily from linked data files as from simple mortality data files, the linked data present opportunities for further useful elaboration. The linkage of birth and death records to provide comprehensive statistics of perinatal and infant mortality is a common example. Linked data may also improve understanding and knowledge of causes of death associated with a large number of procedures and events. Case fatality rates are usually based on unlinked hospital admission and discharge data. In such statistics, only deaths occurring during the particular hospital admission can be recorded. Using linked data, statistics can be extended to include all mortality of patients discharged during a specified time period, thus allowing a more realistic interpretation of hospital in-patient outcome. An appropriate table would show deaths according to the diagnoses recorded for the reference discharge from hospital.

Linkage of death records to earlier social and occupational data provides an invaluable source of information. Examples include the long-term follow-up studies of doctors' smoking habits (Doll and Peto 1976) and the linkage of census data with vital records (Fox and Goldblatt 1982). In such studies mortality rates would be calculated for different social factors and for various diagnostic categories. For example, comparisons of mortality from cancer might be made between smokers and non-smokers, residents of inner city areas and those living in the suburbs, agricultural workers and factory workers, etc.

Health outcome

The single most important tool in the use of linked data is a measure of outcome. It was primarily in order to document the course and outcome of illness and health care that record linkage was developed. Measurements of outcome can include recovery, relapse, readmission, or death.

In many epidemiological and clinical investigations where outcome is contrasted between two or more groups, there may not be sufficient numbers to demonstrate relationships, events may not necessarily occur in an orderly way, and the period of follow-up may be too short. The evident technical and managerial difficulties in this type of research, together with the high cost, have led to increasing use of existing information on which to conduct studies of outcome rather than the mounting of large studies concerned with only one hypothesis. The 'historical prospective' or 'longitudinal cohort' study collects data concerning various measures of outcome in individuals.

Simple cohort analysis

For purposes of comparing the overall achievement of various hospitals, techniques, or even consultants, complication rates based on events occurring within the hospital admission can be biased by the length of time that the patient is in hospital. This was amply illustrated by Acheson (1968) when comparing the outcome of hospital admission with fractured femur in two different hospitals. Initial analysis indicated that the mortality rates at one hospital were much higher than those at another but the duration of stay at that first hospital was much longer than at the second. When linked files were used it was found that the mortality rate within a given time period was identical for each hospital.

A cumulative file of time-sequenced linked data can be analysed in a number of ways to show various types of outcome. The simplest approach is to select a cohort (a group of individuals having one or more characteristics in common), follow all its members for some predetermined period of time, and ascertain what has happened to them by the end of the period. For clinical applications, such as comparisons of procedures, this method may be adequate and practicable. Individuals may be followed for some fixed period, such as one year, after an initial episode involving a specified diagnosis or procedure. The number of those individuals for whom an outcome (such as recurrence) is recorded is then obtained. If required, cohorts from successive time periods may be compared, although interpretation of results is not always easy for conditions where the incidence, mortality, or hospital policy has independently changed over time.

The limitations of such studies concern their interpretation. They are eminently suited to comparison of two groups that are known to be initially comparable in various ways (for example, age/sex distribution, social class background, and marital status). But they are only able to produce valid data concerning the shortest period of follow-up for any one member of the cohort. Thus, no allowance is made for the fact that people are discharged from care over a period of time rather than at a single starting point. Nor is allowance made for the fact that some may die or move away from the reporting network before the follow-up period is complete.

For many purposes it is impracticable to assemble a sufficiently large cohort in which all members are followed for the same period. For larger studies and for those entailing long periods of follow-up, it is necessary to include all individuals satisfying the criteria for entry and correct for varying periods at risk. Further corrections must be made for cohort losses from migration out of the reporting area.

Cohort life table analysis

The cohort life table method provides estimates of risk in the form of probabilities of events occurring within defined time periods following an initial event, such as being discharged from hospital. It has the advantage of being able to make use of all relevant available data on the file.

(1) *Estimation of the risk of a given outcome*. Suppose we are interested in all persons having certain characteristics C and wish to estimate their probability of having an outcome event E (which may be readmission for or death from a specified cause). Let i be the date on which criteria C were satisfied, a be the age of the patient at i, s be the sex of the patient, j be the date on which the patient experienced either outcome E or death from any other cause, and n be the duration of the data file (i.e. the length of time from the start of the data file to the date of the last entry on the data file).

For each patient satisfying criteria C:

If event E occurs subsequent to C, calculate $k = j - i$ and add 1 to the matrix B_{kas}.

If event E does not occur, but death from another cause occurs, compute $k = j - i$ and add 1 to the matrix D_{kas}.

If neither event E nor death from other causes occurs, compute the length of time k from i to the end of the data file and add 1 to the matrix Y_{kas}.

Compute matrix R_{kas} such that

$$r_{kas} = 1 - \sum_{h=1}^{k} m_{has} \tag{3.1}$$

where M_{has} is a matrix of constants which estimates the proportion of the population at risk migrating out of the area in month h.

Then derive matrix T_{uas} where

$$t_{uas} = \sum_{h=u}^{n} [(r_{uas} \cdot y_{has}) + d_{has} + b_{has}] \tag{3.2}$$

which gives the population followed to u months after event C, and who have not experienced outcome E by this time.

The statistic

$$\frac{b_v}{t_v} = \frac{\sum_{a,s} b_{vas}}{\sum_{a,s} t_{vas}}$$

gives the proportion of vulnerable patients with C who will have E in month v, and

$$P_u = \left[1 - \prod_{v=1}^{u} \left(1 - \frac{b_v}{t_v} \right) \right] \tag{3.3}$$

gives the proportion of all patients with C who will have E sometime before the end of u months.

(2) *Applications of cohort life table analysis.* Cohort life table analysis is probably the most useful technique of estimating outcome from linked data. It makes maximum use of the available information. Cohort life tables may be used in several ways to quantify the risk of an event occurring, such as readmission to hospital or death. For example, the outcome following discharge from hospital can be compared in different populations and in different time periods. Table 3.1 shows the probabilities of readmission and death for males, aged 35 and over, discharged with ischaemic heart disease, for follow-up periods up to four years. It can be seen that although the readmission rate is substantially higher for residents of Area 1, the mortality rates are essentially identical.

This use of linked data may be extended. It would be possible to calculate the probability of having a predetermined number of readmissions in a period or, after having had a stated number of admissions, the probability of having a further readmission within a given time. The type of condition

Table 3.1 Cumulative probability of readmission or death, after specified intervals, of males aged 35 years and over having been discharged (1967–1970) with a diagnosis of ischaemic heart disease

	Probability (per cent)			
	Readmission (any condition)		Mortality (all causes)	
	Area 1	Area 2	Area 1	Area 2
Starting Cohort n =	1393	956	1393	956
Interval				
1 week	16.3%	11.8%	1.8%	1.0%
1 month	19.6%	14.3%	4.6%	4.5%
2 months	23.8%	16.4%	7.1%	6.5%
3 months	27.4%	19.1%	8.8%	8.5%
6 months	32.6%	22.4%	13.2%	12.2%
1 year	39.0%	28.2%	19.1%	17.6%
2 years	45.5%	34.4%	25.5%	24.4%
3 years	47.4%	36.6%	29.4%	27.7%

Source: Oxford Record Linkage Study

associated with the readmission could be specified, such as a readmission with the same diagnosis as the index admission. The risk of readmission could be obtained in relation to specified periods of stay in the index admission, or in relation to the likelihood of accumulating some defined period of total stay across all admissions. The chance of changing or retaining the same diagnosis, specialty, or consultant could be measured. These and other approaches could be related to various administrative and demographic characteristics of patients, hospitalization, and death, provided that sufficient numbers of individuals were available for each cohort to yield stable probabilities. The technique has epidemiological and clinical uses as well as applications in health services management and research. An example of a clinical research application is given in Chapter 6.

(3) *Interpretation of cohort life tables.* Care must be exercised in interpreting cohort life table analyses. The resulting probabilities are for each complete interval of time from the start of the follow-up period, and are measures of the risk of the given event (readmission or death) occurring to an individual in the period. The probability of readmission represents the chance an individual has of at least one further inpatient admission following the index discharge, and the probability of death is the chance that an individual will die following the index discharge. By subtracting the probability for the preceding interval, the additional risk of readmission or death, in a stated interval, of individuals who have not been readmitted or died previously can be extracted.

Two questions naturally present themselves when examining cohort life table data:

(*a*) Are there periods of increased or decreased risk during the survival of a single given cohort?

(*b*) Are any two particular cohorts behaving in the same way as each other?

Slightly different analytical methods are needed to answer these two questions and they will be examined separately.

In general the results will be grouped (observations are only made at certain times and the exact time of death or readmission is unlikely to be presented). In addition, for each successive time period the denominator will be altered to take account of the fact that some individuals are lost to follow-up through migration or death. These two aspects introduce several difficulties and the interested reader is referred to Gross and Clark (1975).

(*a*) *Behaviour of a single cohort.* If the probability of the outcome under consideration is constant over time (i.e. there are no periods of increased or decreased risk), then the cohort should decay in a logarithmic manner. This hypothesis can be tested for grouped data by estimating the 'force of mortality' λ and generating the expected numbers (E) of those having the outcome in each interval. Such numbers should be compared with the observed numbers (O) using the fact that $\Sigma(O - E)^2/E$ approximates to a chi-squared statistic.

The parameter λ can be estimated thus: Let there be k intervals of z_l $(l \leqslant k)$ at which observations are made of both the numbers having the outcome and the numbers lost by censoring (these times could be one week, one month, one year, etc.). Let the number of individuals having the outcome in the ith interval be d_i and the number of censorings c_i. Then

$$\lambda \simeq \frac{\sum\limits_{i=1}^{k} d_i}{\sum\limits_{i=1}^{k} (d_i + c_i) z_i} \tag{3.4}$$

to a first approximation. This actually underestimates λ because the exact times of death (or readmission) and censoring are unknown and are taken to be at the end of each interval. Although it is unlikely that, on average, each 'death' occurred at half the time period, this is probably the most accurate approximation. This, in effect, would double the estimate of λ.

The expected numbers of 'deaths' in each interval i can now be estimated as $E_i = t_i (1 - e^{-\lambda z_i})$, where z_i is the length of the ith interval and t_i is the number of persons at risk at the start of the interval. Goodness of fit to exponentiality can then be measured by a χ^2 test on $(k - 2)$ degrees of freedom.

An alternative and graphical examination of the validity of this hypothesis of exponentiality, valid if the degree of censoring is small, is to plot the $\log_e$ of the number remaining in the cohort against time. The result should be a straight line with a negative gradient proportional to λ. Any great departure from linearity would suggest periods of differing risk.

(*b*) *Comparison of two cohorts*. The appropriate test to assess whether two cohorts are behaving in a similar way is Gehan's non-parametric test for censored, grouped data (Gehan 1965).

Let f_{io} = No. of deaths in ith interval for oth cohort ($o = 1, 2$),
    c_{io} = No. of censored observations in ith interval for oth cohort,

and $F_{io} = \sum\limits_{r=1}^{i} f_{ro}$ $(i = 1, k)$ $\tag{3.5}$

Calculate W as

$$W = \sum\limits_{i=1}^{k} [(f_{i1} + c_{i1}) F_{(i-1)2} - (f_{i2} + c_{i2}) F_{(i-1)1}] \tag{3.6}$$

Gehan (1965) shows that if the two cohorts are behaving in the same way then W has a normal distribution with mean zero. Formulas for the variance are given in his paper. Other methods are proposed by Mantel (1967), Efron (1967), Breslow (1970, 1974), and Peto and Peto (1972).

Statistical associations between diseases in individuals

The literature on associations between diseases is now very extensive. In nearly all fields of medicine, instances are cited of relationships between diagnoses. Many are sufficiently well known to be recognized complications of diseases, such as the cardiovascular pathologies associated with diabetes mellitus. Others, like a reduced risk of cancers in schizophrenia, have been reported on the basis of clinical experience, and some have been postulated on the basis of similar demographic patterns in incidence rates, for example, cancer of the breast and cancer of the colon.

The importance of such relationships between diseases is self-evident, both for monitoring individuals at increased risk and increasing knowledge of common features of aetiology. Reduced risks are equally of fundamental importance and may imply some form of biological protective mechanism or suggest negative associations with demographic factors.

Although some associations occur with sufficient frequency to become commonplace clinical observations, there are difficulties in estimating the degree of association from small samples of presenting cases. Similarly, correlations between incidence rates do not necessarily imply that the conditions are statistically associated within individuals. Special surveys to measure association of diseases in individuals are costly and normally limited to preselected sets of conditions. The choice of controls is difficult and the results subject to bias. Rare but statistically important associations are unlikely to be detected by such methods. A large comprehensive linked data file of individual morbidity histories provides a remarkably powerful resource, both for measuring the strength of known or suspected statistical associations in individuals, and for searching for unsuspected associations.

Analysis of disease associations in linked data

The questions posed by this type of analysis of linked data are: (1) What is the risk of event A following event B, or of B following A, or of A and B occurring together? and (2) Is the risk sufficiently higher or lower than that expected by chance to indicate an unexplained aetiological connection between A and B or between them and some other variable?

The objective is to ascertain the number of individuals with A and B, and compare it with a valid estimate of the number of cases that would be expected to have occurred by chance. The ratio of observed to expected cases constitutes a relative risk, which can be tested for statistical significance.

An appropriate method must take into account the actual periods over which all the individuals in the sample are at risk of contracting each of the conditions. The period at risk will be limited by removal from the survey population, by death, or by emigration, while the period over which an individual may acquire the second disease of the pair will be limited to the time available from the point at which the first disease occurred. If there is a time trend in the incidence of the conditions, such as a secular or cyclic variation, the actual risk will vary with calendar time. Clearly, the actual risk of acquiring many of the conditions will also be related to the sex and age of the individuals and probably other variables as well. Furthermore, if the period of follow-up is at all long, the risk may change with the ageing of the individual.

Preparation of the data file

The linked data are likely to contain complex series of episodic records, more than one of which refers to the same disease. For the purpose of this analysis, only the first mentioned occurrence will be relevant, so that readmission to hospital for the same condition, for example, should not be counted in the calculation of either observed or expected numbers. Table 3.2 shows that this permits a simplification of the data. The case illustrated would contribute one observation to each of the sequences: A followed by C, A followed by D, B followed by C, B followed by D, C followed by D and A simultaneously with B. If every episodic record contains several diagnoses, only the first mention of each condition under study should be recorded in the simplified file.

The difficulty of coping with the fact that the age of each individual changes over time is simplified by referring each individual to his year of birth, and hence to his birth cohort c.

Table 3.2 Simplification of linked records to facilitate analysis of time-ordered sequences of diagnosis

Complete Record		Simplified Record	
Event No.	Diagnosis	Event No.	Diagnosis
1	A,B	1	A,B
2	A		
3	A,B		
4	C,B	2	C
5	B		
6	A,D	3	D
7	A,B		

Calculation of expected numbers

The computation of expected numbers of individuals with diagnosis D_1 followed by diagnosis D_2, should be commenced by scanning the simplified set of person records for all those with a diagnosis D_1. For each such individual, the date i of discharge with D_1 and the date j of the end of the period at risk (death or the end of the follow-up period) are identified and 1 added to the appropriate cell of the matrix R_{ijcs}, where c is the appropriate birth cohort group and s is the sex of the individual; i and j are measured in months from the start of the complete linked data file.

Simultaneously, the data in the simplified file should be scanned to identify all persons with diagnosis D_2. For each such individual, the month d of the relevant admission is read and 1 added to the appropriate cell of the matrix Q_{dcs}. Thus any individual in the population at risk (i.e. cohort c, sex s, resident in the area in month d) will have a probability of admission with diagnosis D_2 in month d given by q_{dcs}/n_{dcs}, where n_{dcs} is the number of persons of cohort c and sex s resident in the area in month d.

Now consider the same individual in month $(d + 1)$. We no longer know whether or not he was still resident in the area, but estimate that, of the population at risk, a proportion m_{gcs} of cohort c and sex s emigrated from the area in month g. The probability of the individual being admitted in month $(d + 1)$ with D_2 therefore will be:

$$(1 - m_{(d+1)cs}) \, q_{(d+1)cs}/n_{(d+1)cs} \tag{3.7}$$

The following month it will be:

$$(1 - m_{(d+1)cs}) \, (1 - m_{(d+2)cs}) \, q_{(d+2)cs}/n_{(d+2)cs} \tag{3.8}$$

which approximates to:

$$(1 - m_{(d+1)cs} - m_{(d+2)cs}) \, q_{(d+2)cs}/n_{(d+2)cs} \tag{3.9}$$

Thus, for any individual discharged with D_1 in month i and period at risk $(j - i)$, the overall probability of subsequent admission with D_2 will be approximately:

$$\sum_{h=i+1}^{j} \left[\left(1 - \sum_{t=i+1}^{h} m_{tcs}\right) q_{hcs}/n_{hcs} \right] \tag{3.10}$$

In general, for the population of persons discharged with diagnosis D_1, the total number expected to be admitted with D_2 will be:

$$\sum_{c} \sum_{s} \sum_{j=1}^{a} \sum_{i=1}^{j} \sum_{h=i+1}^{j} \left[r_{ijcs} \left(1 - \sum_{t=i+1}^{h} m_{tcs}\right) q_{hcs}/n_{hcs} \right] \tag{3.11}$$

where a is the last month of follow-up (for example, for an eight-year period $a = 96$).

Each observed number can be considered to be a realization of a Poisson variable, the mean of which is equal to the expected number. Hence the significance of differences of observed from expected values can be based on the usual test. Thus, suppose there were 10 cases observed and 5 expected, it can be calculated that the probability of observing 10 or more cases would be:

Expected distribution of time intervals

In any instance where a significant association in the time-ordered sequence (i.e. D_1 followed by D_2, but not D_2 followed by D_1) has been found significantly more often than expected, then a further factor to be taken into consideration should be the time interval t_{12} from discharge with D_1 to admission with D_2. Theoretically it might be argued that, if there was a significant association D_1 followed by D_2, but no association D_2 followed by D_1, then there is evidence that D_1 may in some way be responsible for the onset of D_2. It is feasible, therefore, that intervals between the two diagnoses might be significantly different from that expected by chance.

To obtain the expected distribution of the time intervals, the distribution of all possible time intervals between every discharge with D_1 and admission with D_2 is calculated after taking account of sex, cohort, mortality, migration, and a changing population base. Thus, if the matrix R_{ijcs} represents the number of individuals of cohort c and sex s, discharged with D_1 in month i and at risk until j, and if Q_{hcs} is the number of individuals of cohort c and sex s, admitted with D_2 in month h, then the number expected to have D_1 followed by D_2 after month i is given by:

$$g_{ijcs} = \sum_{h=i}^{j} \left[r_{ijcs} \left(\frac{q_{hcs}}{n_{hcs}} \right) \left(1 - \sum_{t=i+1}^{h} m_{tcs} \right) \right] \tag{3.12}$$

for all values of i, j, and h such that $i \leqslant h \leqslant j$. Summing over i, j, c, and s the distribution of G_k can then be obtained, where $k = j - i$. This distribution may be compared with that observed.

Simultaneously occurring diseases

Although the case of simultaneously occurring diseases is conceptually simpler than that of sequentially occurring diseases, in practice the calculation of expected numbers is not trivial. Two broad types of disease with differing temporal behaviour must be considered. Acute conditions have relatively clear-cut onsets, short durations, and definite end points; chronic disorders have uncertain points of onset, persist for indefinite periods (often with episodes of acute disturbance), and may be controlled or partially controlled by treatment. Many important chronic illnesses, such as diabetes mellitus and epilepsy, do not necessarily require hospitalization. Yet when a patient with such an illness is admitted for a separate condition his chronic disorder is likely to be reported as a diagnosis also.

The following examples may illustrate the difficulties in establishing whether or not two conditions occur together in individuals more often than would be expected by chance.

(1) Long-standing epileptic; admitted for the first time with acute myocardial infarction.

(2) Admission with acute appendicitis; also found to have pneumococcal meningitis.

(3) Long-standing diabetic; admitted to hospital for stabilization, also recorded as hypertensive.

(4) Admission with intestinal obstruction; found at laparotomy to have cancer of the colon, but also found on examination to have a breast lump which was malignant.

The information required for interpretation is different for each example.

In *Example* (1), it may be important to know whether the risk of myocardial infarction is greater in epileptic than non-epileptic members of the population. Since epileptics can live in the community without ever being admitted to hospital, the denominators are not available from a file that includes only hospital admissions or deaths.

Example (2) is quite a different situation—both the pneumococcal meningitis and the acute appendicitis are conditions of sufficient severity to have warranted hospital admission independently. The risk of the two occurring simultaneously is therefore the product of the risks of each occurring at a particular point in time.

Example (3) refers to two chronic conditions, neither of which independently would necessarily have warranted hospital admission. An assessment of whether the two occur together more often than would be expected by chance is not possible unless it is known (a) how often the two conditions occur together without resulting in hospital admission, and (b) the prevalence of each of the conditions in the community.

Example (4) concerns two conditions present at the same time, one of which was only discovered because of investigation for another condition.

In general, if both diagnoses reported are of independent acute events, such as those of Example (2), the probability of their occurring at the same time k approximates to the probability of having D_1 in month k multiplied by the probability of having D_2 in k.

For a given individual of cohort c and sex s, the probability of D_1 and D_2 together would then be estimated as:

$$\sum_{k=1}^{j} q_{kcs}\, p_{kcs}/(n_{kcs})^2 \tag{3.13}$$

where p_{kcs} and q_{kcs} are the numbers of persons first admitted with diagnoses D_1 and D_2 respectively in month k.

For cohort c and sex s, the number of persons expected to have D_1 and D_2 at the same admission will be:

$$\sum_{k=1}^{j} (q_{kcs}\, p_{kcs}/n_{kcs}) \tag{3.14}$$

and for the whole population, the expectation will be:

$$\sum_{s}\sum_{c.}\sum_{k=1}^{j} q_{kcs}\, p_{kcs}/n_{kcs} \tag{3.15}$$

This calculation appears to be satisfactory where each member of the pair of conditions independently would have required emergency hospital in-patient admission. For example, linked data spanning five years of the Oxford Record Linkage Study yielded one observation of acute tonsillitis and acute appendicitis; 1.2 cases were expected. Acute appendicitis at the same time as concussion was expected 0.3 times and no cases were found.

Where one or both conditions reported simultaneously are chronic, some means of estimating their prevalence must be used. Prevalence data specific for each sex, year of birth group, and time period are not generally available, but it is possible to estimate prevalence from the simplified data based on all persons with two concurrent diagnoses. Thus suppose that among cohort c and sex s there were b_{csk} individuals discharged with any two concurrent diagnoses in month k, and that x_{csk} were noted to have D_1 and y_{csk} to have D_2. Assuming the two conditions to be independent, the expected number in cohort c and sex s to have both diagnoses would be $x_{csk}\, y_{csk}/b_{csk}$. Then the overall number expected to have both diagnoses would be:

$$\sum_{c}\sum_{s}\sum_{k=1}^{j} x_{csk}\, y_{csk}/b_{csk} \tag{3.16}$$

This also yields some expected values very close to those observed for appropriate diagnoses.

The two methods give very different results when applied to inappropriate pairs of conditions so that it would be necessary to specify the method used. Moreover, some pairs of simultaneously occurring conditions seem to give observed numbers far in excess of expectation. For example, in the Oxford Record Linkage Study files, myocardial infarction occurred at the same time as bronchopneumonia in 226 individuals. Both may be acute conditions yet the appropriate method gave an expected value of 8.7. This may be an artefact in that, if death occurs, death certificates often state broncho-pneumonia as the terminal illness without sufficient evidence for the diagnosis. The extent of such a bias can be assessed by eliminating these double events that resulted in death, and analysing only those where the patient survived. If similar results are found for each, then it may be concluded that they are not the result of poor certification.

In practice, the absence of a generally applicable method makes it difficult to deal with simultaneously occurring diagnoses. Hospitalization for an acute condition sometimes leads to diagnosis of a chronic disease which must have preceded it. For example, a laparotomy for acute appendicitis may reveal a bowel cancer. The particular circumstances of each pair of diagnoses must be considered and the cases assigned appropriately. In an individual with two major chronic conditions that would eventually have resulted in hospital admission or death (such as those described in Example (4), where the second condition was diagnosed slightly earlier because of hospitalization for the first condition), it would probably be valid to assume that the two conditions occurred in a time-ordered sequence.

Assessment of possible errors

In linked records systems there is a risk of linkage failure causing either missed or false links. This topic was discussed in detail in Chapter 2 and will not be considered further here. In addition, both systematic and random errors are common in large collections of processed records. They may arise at any stage from the original clinical records through to the subsequent abstraction, coding, and processing procedures. The most common form of data error is likely to be missing information, where either items are lost from records or whole records are omitted. In well-designed data collection systems, with extensive editing and error detection procedures, the extent of such faults should be known, although their complete elimination is unlikely.

For most data items, the number coded as 'not known' or 'not stated' will be available. Some important items, such as diagnoses, are not of this type since the number of entries in a record may be open-ended and it may be impossible to estimate the extent to which under-reporting has occurred. Conversely, after some in-patient examinations the true diagnoses may well be 'not known'. It is important that these should be coded in such a way that they are not confused with codes implying missing information.

Verification of diagnoses

In order to verify results using a large data base we suggest that some validation should take place. Thus for all major diseases under consideration a randomly determined sample of patients on the file should be examined. In the Oxford Record Linkage Study we have attempted to verify the diagnoses by going back to hospital records or using other means at our disposal. Examples of the results using these methods are shown in Table 3.3. In each instance, 100 cases with a particular diagnosis were chosen and clinical notes reviewed. In addition, cancer registry records (themselves derived from pathologists' reports and other hospital records) were used for cases of cancer. From Table 3.3 it can be seen that among the various cancers and non-malignant disorders the proportion of diagnoses correctly coded was very high. The largest source of error was due to erroneous coding. This was especially high in the group of gastric ulcers, largely due to a confusion between the code for gastric ulcer and that for duodenal ulcer.

Table 3.3 Verification of diagnosis in randomly chosen subsets of records (Oxford Record Linkage Study, 1963–70)

	Cancer of ovary (n=100)	Cancer of bladder (n=100)	Benign breast disease (n=100)	Diabetes mellitus (n=100)	Gastric ulcer (n=96)	Pernicious anaemia (n=99)
Diagnosis confirmed by:						
—cancer registry	93	94	—	—	—	—
—hospital records	—	—	98	82	78	80
—death certification only	4	—	—	18	7	12
Diagnosis not confirmed:						
—tentative diagnosis only	2	3	—	—	2	2
—hospital diagnosis wrong	—	1	—	—	—	—
—death certificate inconsistent with other evidence	—	1	—	—	—	—
—coding error	1	1	2	—	9	5
Proportion of diagnoses assumed correct (%)	97	94	98	100	89	93

The significance of errors, even at low overall rates, is highlighted by their effect on analyses of diagnostic associations. For example, consider a patient who is in and out of hospital with schizophrenia (ICD9 295.2). For one hospital admission the code was miskeyed and he appeared to have had a malignant neoplasm (ICD9 195.2) of the abdomen. In any analysis looking for an association or lack of an association between schizophrenia and cancer, this individual would contribute one case in the order schizophrenia followed by cancer. Only an assessment of the records would reveal that this case should not have been included.

We have shown in Table 3.3 that for most diagnoses the amount of error is fairly negligible (less than 10 per cent). There is therefore little need to amend or put an error correction into the estimate of expected numbers. It might be argued that correcting the observed numbers without correcting expected numbers introduces substantial bias. But this tends to cause a decrease in the observed numbers compared to the expected and so, if a significant excess of observed over expected numbers is still shown, the results will be even less likely to have arisen by chance than is shown by the simple probability level calculated.

However, where a significant negative association is found, it may be wise to correct the expected number. Thus, suppose an assessment of 100 persons with D_1 suggested that 89 per cent were correctly coded, and that the estimate for D_2 was 94 per cent, then the number expected to have D_1 followed by D_2 (given by formula 3.11) would be corrected by multiplying by the factors (0.89)(0.94).

Verification of cases apparently showing association between diagnoses should be checked for other types of error, apart from wrong diagnostic codes. Errors in the time-sequencing of events in the linked record may arise, most likely because of omission of episodes occurring before the start of data collection or of services not covered by it. (Incorrect dating of episodes is usually eliminated by prior editing.) In a study of six common cancers based on Oxford linked data covering eight years, 5.4 per cent of all cases were found to have been first diagnosed before the starting date of the file. A further 4.8 per cent had been treated in out-patient services or in hospitals outside the reporting network, prior to the first entry in the linked data. In such cases, the first admission on the file may not mention the cancer, but rather some other condition D. If subsequent admissions mention the cancer, then it will appear, wrongly, that D preceded the cancer. Clearly, this type of error may be an important potential source of bias.

Error rates are likely to be somewhat higher in records contributing to significant associations between diseases because the cases are usually complex and there are likely to be more episodic records for each individual. Moreover, a diagnostic or coding error may have compounding effects by causing a temporal sequence error and, more seriously, spurious associations.

Selection bias

It is well known that studies based on hospital in-patients are subject to several forms of bias arising from the fact that they are selected from the general population and may not be representative of it. Similar considerations apply to studies of mortality, as shown in Chapter 4.

Surveillance bias

Surveillance bias is the increased likelihood of detection of coexisting disease in patients already hospitalized for a particular disease. Repeated examination and detailed investigation of in-patients, particularly those with chronic conditions causing frequent readmission, increase the chance of early detection of other diseases. Indeed, specialist interest may lead to detection and recording of conditions in certain hospitals (for example, teaching hospitals) which would not be noted in others. To some extent this effect can be offset by use of other in-patient records as comparison groups. However, some differences cannot be corrected, and the size of the effect is difficult to estimate. For example, a middle-aged woman having one admission for appendicectomy may be less likely to be recorded as hypertensive than a similar woman admitted repeatedly with cancer of the cervix.

Berkson's bias

Berkson (1946) showed that if there is an independent probability of admission to hospital for each condition, there is also an increased likelihood of more than one condition in a hospitalized individual than in a comparable individual in the community, since the independent probabilities of admission for each condition are compounded. Though difficult to demonstrate with real data, this important source of bias has diminished reliance on results of hospital record studies. Berkson was concerned with the prevalence of concurrent diagnoses in individuals in a hospital population. This form of bias does not apply to temporal sequences of diagnoses in separate admissions of individuals, where the independent rates of admission are taken into account in the calculation of expected numbers. The bias introduced in considering pairs of diagnoses recorded in the same admission has been considered in some detail on pages 67–70.

Significance level

Although the significance of deviations of relative risks from unity can be based on exact probabilities, normal caution should be exercised. Risks based on small numbers should be treated with circumspection and where large numbers of significant results are obtained, a proportion will be due to chance. In these circumstances, levels lower than $p = 0.05$ should be chosen to minimize misleading results.

Case-control studies

One of the many ways in which the data can be analysed in a linked file employs the case-control approach. This can be used as an adjunct to one of the other methods or as the sole method of analysis.

For example, suppose the diagnostic associations method had shown an increased relative risk of skin cancer after hospital admission for hypertension, but that there was no association between skin cancer and subsequent hypertension. One hypothesis would be that a method of treatment of the hypertension might be a causative factor in the genesis of skin cancer. Such a hypothesis could be tested by carrying out a case-note study comparing cases and controls. The cases would obviously be all those persons treated for hypertension who subsequently developed skin cancer. Considerable thought would have to be given to the choice of controls from the linked file. For each case, controls should be chosen, also admitted for hypertension, and of the same age and sex distribution as the cases. In addition they should have survived at least for as long as the index cases but they must not have developed skin cancer. Other factors might be considered for matching, but great care would have to be taken not to over-match. For example, if a particular consultant were to treat all his hypertensive patients in the same way, then the matching on consultant would nullify any real effect that might be present.

A further use of the method can be to delineate characteristics of a given disorder when there are clearly major interacting associations. This is feasible only if the whole population at risk is held on the file. For example, the sudden infant death syndrome is strongly and independently related to social class of the father (the lower the social class the higher the risk), the mother's age (the younger the mother the higher the risk), and number of previous pregnancies (the more the higher the risk). By picking controls matched on these factors, it has been possible to assess whether, over and above them, there is any association between sudden infant death syndrome and such factors as inter-pregnancy interval, breast feeding, and birthweight of the infant (Golding, Limerick, and Macfarlane 1985). Use of random controls has a place in studies where little is known about a condition, but given a large data base there is little theoretical justification for such a procedure when far more sophisticated tools are available.

Genetic and other family studies

Assuming that the available linked data can be used to derive family records, either directly or by means of the family record linking methods described in Chapter 1, several types of analysis can be carried out, depending on the type of question to be answered. Probably the simplest is estimation of the relative risk of some condition occurring in relatives of a proband group in comparison with the risk in unrelated individuals. Here the rates of occurrence in the relatives, classified by relationship, are compared directly with the rates of occurrence in the population from which the families were derived. Adjustment should be made for sex, age, period at risk, and other variables in the usual way and, in general, no special techniques are necessary. An example is the determination of whether relatives of children with retinoblastoma are at higher risk of other cancers than the general population (see Chapter 17).

A further example of this type of study is the determination of whether spouses of psychiatric patients are at greater than expected risk of mental disorder. The expected number must be calculated from an estimate of the number of married pairs in the general population under study. A consideration of the appropriate statistical method for this estimation is given in Hall, Baldwin, and Robertson (1971). These authors point out that the simple assumption that the number of couples in a population is half the sum of the total number of married males and females, is unlikely to be valid and that the calculation is further complicated by the varying periods of observation of the patients. Nevertheless, they were able to devise a formula applicable to available population data which gave a reasonable estimate of the number of couples. Unfortunately, published population data do not provide analogous figures for siblings and parents and, therefore, this form of analysis is not possible unless data on the births of the members of the population are held on the file.

Where the linked record system contains sufficient information to link all the records into families and the resulting file of family records is approximately complete, analysis can be focused on particular types of relationship, such as the risk in siblings. The classic study of Newcombe (1966) linked the British Columbia vital records system with the handicap register. It is reproduced in Chapter 14.

If the birth population is on file and disorders of infancy and childhood in siblings are under consideration, we would suggest a case-control approach similar to that described in Chapter 15. Depending on the type of relationship, suitable controls would be picked. Thus, in the study of anencephaly in the Oxford Record Linkage Study, for each case of anencephalus we were able to pick control children born in the same year, to mothers of the same age, parity, and social class. Siblings of cases and controls were identified from the linked file and linked to their hospital records. Comparison of abnormalities in index siblings and control siblings was used to assess recurrence risks.

References

ACHESON, E. D. (1968). The incidence and prognosis of fracture of the femoral neck. *Unit of Clinical Epidemiology Report* **No. R17** , Oxford University.

BERKSON, J. (1946). Limitations of the application of fourfold table analysis to hospital data. *Biometrics Bulletin* **2** , 47–53.

BRESLOW, N. (1970). A generalized Kruskal-Wallis test for comparing K samples subject to unequal patterns of censorship. *Biometrika* **57** , 579–94.

—— (1974). Covariance analysis of censored survival data. *Biometrics* **30** , 89–99.

DOLL, R. AND PETO, R. (1976). Mortality in relation to smoking: 20 years' observations on male British doctors. *British Medical Journal* **2** , 1525–36.

EFRON, B. (1967). The two sample problem with censored data. *Proceedings of Fifth Berkeley Symposium* **4** , 831–54.

FOX, A. J. AND GOLDBLATT, P. (eds) (1982). *Longitudinal Study: socio-demographic mortality differentials 1971–75*. HMSO Series LS NOI1, Office of Population Censuses and Surveys, London.

GEHAN, E. A. (1965). The generalized Wilcoxon test for comparing arbitrarily singly-censored samples. *Biometrika* **52** , 203–23.

GOLDING, J., LIMERICK, S., AND MACFARLANE, J. (1985). *Sudden Infant Death: Puzzles, Patterns and Problems*. Open Books, Shepton Mallet, and University of Washington Press, Seattle.

GROSS, A. J. AND CLARK, V. A. (1975). *Survival distributions*. Wiley, New York.

HALL, D. J., BALDWIN, J. A., AND ROBERTSON, N. C. (1971). Mental illness in married pairs: problems in estimating incidence. In *Aspects of the epidemiology of mental illness: studies in record linkage* (ed. J. A. Baldwin). Little, Brown, Boston, Mass.

MANTEL, N. (1967). Ranking procedures for arbitrarily restricted observation. *Biometrics* **23** , 65–78.

NEWCOMBE, H. B. (1966). Familial tendencies in diseases of children. *British Journal of Preventive and Social Medicine* **20** , 49–57.
PETO, R. AND PETO, J. (1972). Asymptotically efficient rank invariant test procedures. *Journal of the Royal Statistical Society (A)* **135** , 185–98.
World Health Organisation (1977). *Manual of the international statistical classification of diseases, injuries and causes of death*, 9th revision, 1975, vol. 1. Geneva.

PART II

Applications of linked records

4

Selection and mortality differentials

A. J. FOX, P. O. GOLDBLATT, AND A. M. ADELSTEIN

The Office of Population Censuses and Surveys Longitudinal Study provides reliable mortality data by a much wider range of characteristics than are available from other national sources. Although it is based on only a 1 per cent sample of the population, it broadens the scope of mortality analysis and permits study of changes in relationships using different aspects of the time dimension. Data from this study have made us increasingly aware of the importance of selection to the interpretation and understanding of observed mortality differentials. Here we focus on that aspect of selection called 'health-related mobility', which is associated with the relative health of people acquiring or losing individual characteristics. It is suggested that, for characteristics affected by health-related mobility, mortality differentials would narrow or widen with increased duration of follow-up. On the basis of this argument the contribution of health-related mobility to mortality differentials by economic position and social class, to regional differentials, and to family and household differentials is investigated. Selection can thus be shown to operate when people change economic position, when they migrate, or when they change marital status. While the effects of these selection processes can be shown to contribute to social class gradients they do not explain regional differentials and contribute only to a limited degree to differentials by marital status. Differentials by household circumstances also reflect the product of selection processes.

The Office of Population Censuses and Surveys Longitudinal Study brings together for analysis information obtained from several vital records. It can currently be analysed as a prospective study of mortality, incidence of cancer and survival, fertility, migration, or widowerhood. Although limited in size— it is based on a 1 per cent sample of the population—it has three main advantages over traditional data sources. The most obvious is that it overcomes bias between numerator and denominator that—for example, in occupational mortality statistics—may be large enough to call the whole analysis into question. Secondly, it extends the scope of analysis because OPCS can now calculate rates according to all census characteristics or combinations of characteristics. Finally, it provides measures of the change in relationships with time.

Reprinted from *Journal of Epidemiology and Community Health* **36**, 69–79 (1982).

Traditionally, national mortality statistics have been used to study age, period, and cohort effects. Recently, interest in time has broadened as epidemiologists in particular have learnt to use prospective studies—bringing together exposure and subsequent mortality records—to look at other time-related effects. Studies have been based on cohorts entering or leaving employment at the same time. Others have looked at the effect of duration of exposure (as distinct from level of exposure) and at different time relationships between exposure and effect. Prospective studies create the potential for these analyses because they incorporate historical information preceding follow-up and permit analysis by duration of follow-up.

Two aspects of time are central to the following discussion of selection and mortality differentials. These are the timing of changes in characteristics and the duration of follow-up. The OPCS Longitudinal Study has enabled us to show this influence on mortality differentials for a broad range of characteristics.

Selection and health-related mobility

Over the past few years we have been analysing mortality from this study systematically by each census characteristic—looking in particular at employment, household and family structure, housing, education, area of residence, and migration history. The objectives of our analysis were, firstly, to describe the relationships observed and then to try to understand and interpret them. As other researchers in the field we found ourselves continually dismissing some relationships as 'due to selection', while others were considered important because they were thought to be a consequence of the circumstances implied by the characteristics in question. It was unsatisfactory that such judgments were made on a subjective basis, and we consequently sought ways of measuring the extent to which mortality was influenced by selection.

The term selection is used widely in epidemiology, particularly in occupational studies that invariably refer to the 'healthy worker effect' and in studies of mortality by marital condition and those of migrants. Never quantified, it is usually justified by comments such as: 'sick people do not seek employment', 'sick people do not marry', or 'sick people do not migrate'. By describing what happens at the extreme the authors have shown that the problem exists.

Sometimes the term is used in a broad sense to describe the social circumstances, behaviour, or way of life of people with the characteristic in question or even their genetic characteristics. In this way the epidemiologist is trying to indicate that the observed mortality does not simply reflect the environment implied by the characteristic in question but also several other factors.

This chapter focuses on that aspect of selection that we call *health-related mobility*. It describes the artificial raising or lowering of the average health of people with a particular characteristic associated with the processes by which that characteristic is acquired or lost. The mortality of a population with that characteristic is affected by health-related mobility if the health of people acquiring or losing the characteristic differs systematically from others with the characteristic.

From this description of the process it should be seen that this aspect of selection operates from the time the characteristic is acquired continuously to the time the characteristic is lost. The description accounts for selection on the basis of ill health as well as the more commonly described health selection. Also it carries no implication as to who in selecting; the process may reflect individual decisions or institutional decisions.

To understand what would be expected to happen to mortality as a result of health-related mobility, consider health as a two-state process—people are either healthy or sick. Health selection is said to operate when the prevalence of sickness is lower among those people acquiring the characteristic and higher among those people losing the characteristic than among all people with the characteristic. When ill-health selection operates, the opposite pattern will be seen. After follow-up has begun the prevalence of sickness will reflect the prevalence of sickness at the time of selection and the balance between new cases of sickness and recovery, emigration, or death of old cases. If the prevalence of sickness was artificially raised (or lowered) by selection, through the inclusion (or exclusion) of sick people, then the number of new cases of sickness occurring after follow-up has begun can be shown to be lower (higher) than the number of sick people who recover, emigrate, or die. The prevalence of sickness would therefore reduce (or rise) with increased follow-up.

To predict what happens to mortality it would be necessary to know how sickness was related to mortality. In this context it is useful to distinguish between chronic sickness and acute sickness. Chronic sickness would be expected to lead to relatively high mortality with a gradual decline with increased duration of follow-up which reflects the low recovery rate. On the other hand, acute sickness would probably lead to a very high mortality in the short term (reflecting the acuteness of the sickness) which would be followed by a sharp drop in relative mortality (reflecting the high short-term recovery rate) and a gradual decline from a high level in the medium and longer term (reflecting the mortality of those who become chronic sick).

Given these relationships between sickness and mortality and the above comments about the change in prevalence of sickness with increased duration of follow-up, relative mortality would also change with time until it approached the level implied by the steady state prevalence of sickness.

If, on the other hand, the prevalence of sickness at the start of follow-up was near to the steady state balance between new cases and exclusions, the prevalence of sickness and relative mortality would be expected to remain stable over time.

The clearest single illustration of the importance of the distinction between acute and chronic sickness and of the change in differential for health and ill-health selection is a diagram presented in an early report based on the OPCS Longitudinal Study. Figure 4.1 (taken from Goldblatt and Fox 1979) shows the changes in relative mortality for residents and non-residents in private and non-private households by year of death. Since the sample was classified by their characteristics in 1971 the year of death is equivalent to the duration of follow-up.

Consider firstly those people who were enumerated in their private household of usual residence. By definition this group would exclude people such as those who were sick and disabled and were, at the time of census, resident in hospitals and those who were visiting hospitals for acute medical problems. With increased duration of follow-up people in this group would,

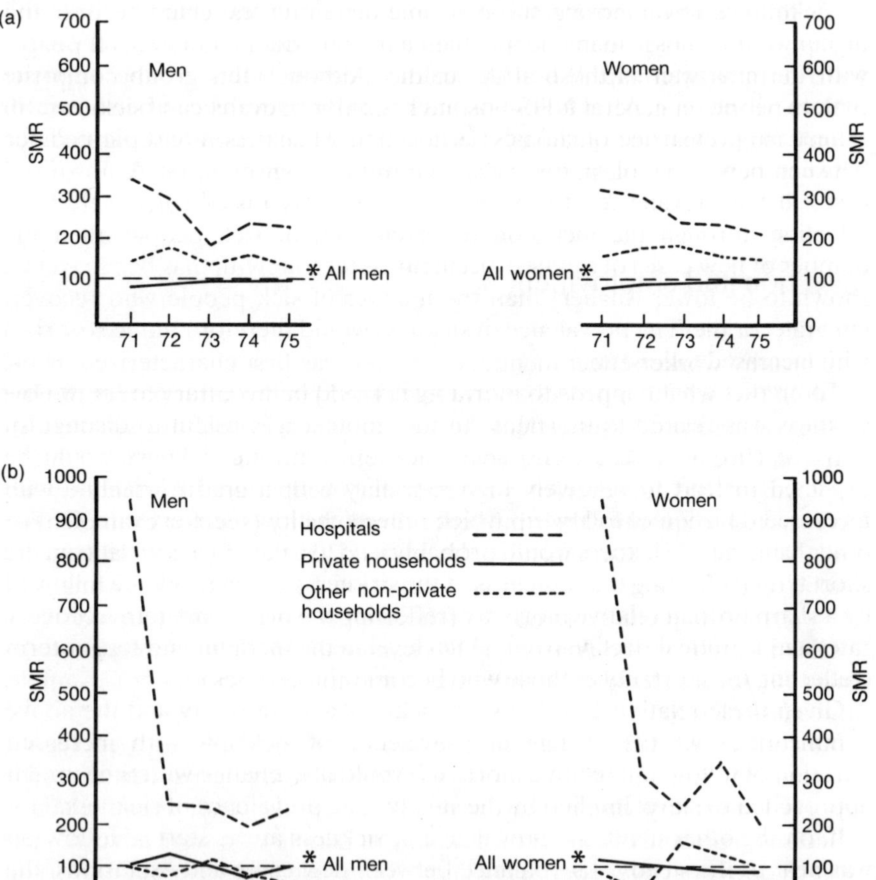

Fig. 4.1 Mortality of (a) residents and (b) non-residents in private and non-private households by year of death. * Longitudinal study population.

however, become sick and would go into various types of institution. Therefore our model suggests that initial mortality for the group would be low and that it would rise with increased duration of follow-up. This is the pattern observed in Figure 4.1, which suggests that after three or four years the selection effect would have worn off.

Now consider those people enumerated by the census in their hospital of usual residence. For such a group of chronically sick people our model suggests that relative mortality would be raised and would decline gradually with increased duration of follow-up. The decline observed for this group in the figure is gradual, and even after five years the SMR is still above 200.

The main group of interest in the bottom half of the figure comprises people enumerated in hospitals but usually resident elsewhere. The figure shows the expected sharp decline in relative mortality between 1971 and 1972 followed by a more gradual decline thereafter: expected because this group would include many people with a high probability of recovery in the short-term as well as those about to die. Although this group comprises mainly people in general NHS hospitals, similar patterns can be shown for people in psychiatric or in non-NHS hospitals when these are separated out. The numbers of people in these latter groups are, however, small.

Selection and employment

The healthy worker effect mentioned earlier was first characterized as the result of two selection processes by Ogle (1885) in the latter part of the last century. These processes reflect the way in which people are selected for work and the way they move from one job to another as well as out of employment. It is, however, only recently that occupational mortality studies have started to quantify their influence on mortality (see, for example, Fox and Collier 1976). Certainly by the mid-1970s there was wide concern among epidemiologists engaged in occupational studies that they might be missing important clues as a result of the healthy worker effect. In a series of articles in 1976 in the *Journal of Occupational Medicine* several authors called for standardized mortality rates for employed people—for example, Goldsmith (1975).

As well as providing data on the mortality of employed people, the Longitudinal Study illustrates an important way study design can affect the observed mortality differentials, irrespective of the underlying selection processes. This study is based on a sample of people who were alive in April 1971; it does not cover complete cohorts of people who went out to work or who left employment. Since the underlying selection processes reflect selection at entry into work and selection at the time of leaving work, it would be necessary to observe new entrant and recent leavers to obtain proper estimates of the influences of these processes. Instead, the census design gives a sample of people who were working at the time of the census, some of whom would have been in their jobs for a long time.

The models proposed in the previous section suggest that the prevalence of sickness in population identified by a cross-sectional census would be closer to the steady state prevalence for people with the particular characteristic than the prevalence of a group of people who recently acquired or lost the characteristic. While this means, however, that the effects of selection will be diminished by the cross-sectional census design, it also implies that estimates of their magnitude may be conservative. These comments apply not just to employed men (for whom men who recently obtained employment would show more pronounced selection effects than all employed men) but also for married men (newly married compared with all married) and for migrants (recent migrants compared with all migrants).

Figure 4.2 gives the SMRs for men and women who were employed the week before the 1971 census by age and year of death. For each group, the SMRs are lower in 1971 and rise progressively in subsequent years. There is the suggestion that the effects are greater for women and for men after normal retirement age than they are for men aged 15–64. Despite the census design—that is, the conservative estimate of selection—the effects are still evident in 1975, after nearly five years of follow-up.

In a more detailed analysis of these data (Fox and Goldblatt 1982) we have shown that for both sexes, health-related mobility was strongest for respiratory diseases and circulatory diseases. For both sexes there was also a clear gradient for malignant neoplasms but no gradient for either sex for accidents and violence. This pattern is identical with that noted earlier in a study of vinyl chloride workers (Fox and Collier 1976).

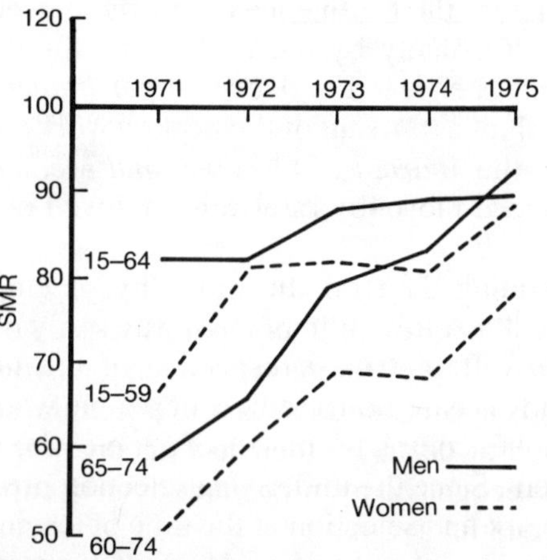

Fig. 4.2 Mortality of employed men and women by year of death and age.

Selection and social class

As the Registrar-General's classification of social classes is based on occupation and as occupational descriptions for the 1971 census varied in quality according to the person's economic position, it is to be expected from the previous section that the social class gradients in the longitudinal study would also be affected by selection.

Figure 4.3 gives the observed gradients based on mortality in 1971–5. There are two major differences between the pattern observed here and that presented in the most recent decennial supplement on occupational mortality (Office of Population Censuses and Surveys 1978). Firstly, the gradient between men in the various classes is flatter in this figure than in that previously published. Secondly, the group who were unoccupied at the

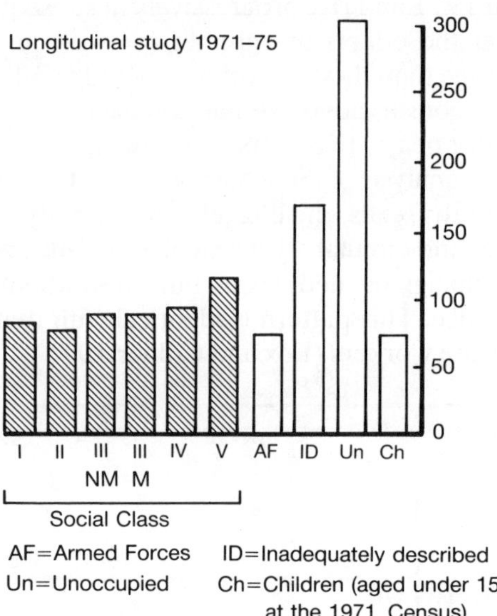

Fig. 4.3 SMRs for men aged 15–64 by social class.

census have an SMR of over 300, compared with one of under 50 in the decennial supplement. The differences between the two methods are easy to understand once it is appreciated that in the decennial supplement analysis, people are classified at death by their last occupation before death. Since 'unoccupied' is not recognized by informants as an occupational category it is rarely recorded at death.

In an attempt to assess the extent to which this explained the differences between the two methods we allocated the unoccupied into social classes (see, for example, Fox 1980). An alternative approach would have been to consider the categorization unoccupied as a selection process out of social classes. If the selection out of work acts disproportionately for each class then the gradient observed between the classes would be expected to change with duration of follow-up. Table 4.1 shows that in 1971–2 there was little difference in mortality between men in the various social classes. By 1974–5, however, a clear gradient had appeared—from 60 for social class I to 124 for social class V. It remains to be seen from further follow-up whether this gradient represents the steady state difference between social class I and social class V. Nevertheless, it probably represents a more accurate estimate of the underlying difference than the average figure for 1971–5.

Table 4.1   Social class gradients by year of death

Social class		1971–2		1973		1974–5	
I							
Obs		63		31		42	
	SMR		105		89		60
Exp*		59.9		34.8		69.5	
II-IV							
Obs		1039		618		1235	
	SMR		86		91		92
Exp*		1201.6		681.3		1335.4	
V							
Obs		137		73		153	
	SMR		111		110		124
Exp*		123.6		66.3		123.7	

* Expected deaths based on age-specific rates for all men in the Longitudinal Study in five-year age groups.

Selection and migration

The effects of migration on geographic variations in mortality have also caused concern for some time. Welton in 1872 highlighted three groups of migrants that might influence geographic variations. One group comprised generally stronger individuals than those who remained and would be expected to raise the mortality of the area they left. The other two groups migrated for reasons of ill health; the first returned to the area from when they came in order to die and the second comprised long-standing town dwellers who moved to the countryside to convalesce.

The mortality of recent migrants can be derived from the longitudinal study because the 1971 census included questions about addresses in 1966, 1970, and 1971. A relatively simple analysis by distance migrated points to

Table 4.2 Mortality of people migrating in the five years before census by distance moved and time since migrating*

Distance migrated		Men		Women	
		Time between migration and census		Time between migration and census	
		< 1 year	≥ 1 year	< 1 year	≥ 1 year
Moved:					
Within county	Obs	657	1659	698	1846
	SMR	116	104	115	111
	Exp	564.7	1602.4	606.6	1656.2
Between counties in a region	Obs	85	259	94	234
	SMR	92	90	106	90
	Exp	92.2	286.5	88.7	259.0
Between regions	Obs	85	252	84	234
	SMR	88	90	96	91
	Exp	97.0	280.7	87.6	257.7
All migrants within England and Wales	Obs	827	2170	876	2314
	SMR	110	100	112	106
	Exp	753.9	2169.5	782.9	2172.8

* Compared with all those in England and Wales as standard.

at least two selection processes (Table 4.2). Migrants who had moved only short distances tended to have high mortality. There is little difference in the SMRs between men and women or between migrants who moved in the year before the census and those who had moved earlier. If these data are separated by age they indicate that the high mortality among local migrants reflects high mortality for men aged 75 and over and women aged over 45. The SMRs rise progressively with age for these groups. The low mortality for regional migrants is concentrated in the 45–74 age range for both men and women.

The SMRs are above 100 because migrants as a whole are dominated by local migrants. The table, however, also shows that if our interest is in geographic, in particular regional, patterns then it is essential that the under-lying pattern of migration be understood. The arguments in the preceding sections suggest that if these differentials reflect health-rated selection among the different groups of migrants then the differentials would be expected to change with increased duration of follow-up. Table 4.3 shows that, for men in particular, the change in SMRs with time would be consistent with local migrants including a few people whose move was motivated by their ill health and regional migrants excluding people whose ill health prevented them moving. Also of importance in Table 4.3 is the pronounced change in SMR for men and women who moved between counties within a region of 12 months before the census. The change in SMR suggests that although this

Table 4.3 Mortality of people migrating in the five years before census by distance migrated, time since migration, and calender period*

	Time between migration and census			
	< 1 year		> 1 year	
	Year of death		Year of death	
	1971–2	1973–5	1971–2	1973–5
Men				
Moved:				
Within county	127	110	107	102
Between counties in a region	120	77	94	88
Between regions	77	94	82	94
Women				
Moved:				
Within county	111	117	119	108
Between counties in a region	136	89	89	91
Between regions	95	96	98	87

* Compared with all those in England and Wales as standard.

group comprises mainly healthy migrants, it may also include a disproportionate number of people whose move was related to an episode of acute sickness. The mortality of those migrants who moved before 1970 is similar to that of regional migrants, suggesting that the effect of the acute sickness is no longer evident among such migrants more than a year after they moved.

Analysis of mortality by the full range of census characteristics has indicated that household characteristics to a great extent describe the need for care and support. Apart from people living in homes for the aged and hospitals there are those who, unable to live on their own, are supported by friends or other relations in private households. Table 4.4 describes the mortality of recent migrants according to household type, when they migrated, and year of death. It contrasts those people who moved recently, probably for support reasons, with those who had moved sometime earlier.

The pattern of change with time, both in relation to when migration took place and in duration of follow-up, is clearest for residents in hospitals. For both sexes the SMRs are higher for recent migrants than for those whose address had not changed in the previous five years. At the same time, irrespective of when the move was made, mortality declined with duration of follow-up.

The SMRs for residents in homes for the aged were not as high as those for residents in hospitals nor were the time changes so clear. For men who moved into these places more than one year before the census there was a clear suggestion of decline in SMR with duration of follow-up and time of migration. Nevertheless, the SMR in 1971–2 for men who moved more than six months but under one year before the census was lower than would have been expected. For women the SMR for this latter group was also lower than might have been expected and that for 1973–5 for non-migrants in homes was higher than expected. The only clear time change was for those who had moved between one year and five years before the census.

Since visitors to hospitals comprise primarily acutely sick people it is not surprising that their mortality does not appear to be related to when the move took place as much as to duration of follow-up. Whereas for the two groups above, the move recorded by the census included the move to the particular institution, for visitors it simply reflected a change of address. There is the suggestion from this table that recent migrants who were visiting hospitals did so for less serious health reasons than did those who moved more than a year before the census and those who had not moved. Nevertheless, irrespective of whether or not they had moved and when they had moved, visitors to hospitals showed the characteristic pattern of mortality expected for the acutely sick, already shown by the bottom half of Fig. 4.1.

The patterns for people in private households are especially interesting. Table 4.4 suggests how the relationship between support and mortality extends into private households. Those people who were living in multi-person households but were not part of a family included many who, after their family dissolved (through the marriage of children and death of a spouse), were not able to maintain themselves independently and consequently moved in with their children's family or with other relatives. This group can be seen to have higher mortality than others in private households, which comprise mainly those living alone or as married couples. Interestingly, for both these groups, relative mortality rose with duration of follow-up. This suggests that for both groups the selection-out of the sick into homes and hospitals is the dominant factor affecting short-term changes in mortality.

Selection and region

One of the reasons for interest in the mortality of migrants has been to assess whether the historic pattern of migration from regions in the north and west of England and Wales to the southern and eastern regions may explain regional mortality gradients that for more than 100 years have seen high mortality in the north and west and low mortality in the south and east. This was the background to the analysis of Welton (1872) in the last century and of Stevenson (1913) and Bradford Hill (1925) in the early part of this century.

Table 4.5 describes the mortality of people who migrated in the five years before the 1971 census by region of residence in 1966 and in 1971. Because the numbers are relatively small, the table is presented to describe movement

Table 4.4 Mortality by type of household migration in the five years before census, time since migration, and calender period

Type of household		Time between migration* and census: <1 year, Year of death 1971–2	<1 year, Year of death 1973–5	≥1 year, Year of death 1971–2	≥1 year, Year of death 1973–5	Non-migrants*, Year of death 1971–2	Non-migrants*, Year of death 1973–5
Men							
Non-private Homes	Obs	34	45	69	55	52	52
	Exp**	22.1	27.8	34.0	37.5	31.9	42.9
	SMR	154	162	203	147	163	121
Resident in hospitals	Obs	23	14	24	16	66	71
	Exp**	4.1	3.5	8.2	7.5	26.3	36.9
	SMR	561	400	293	213	251	192
Visitors in hospitals	Obs	9	6	20	15	148	74
	Exp**	2.8	3.1	5.0	6.2	24.9	33.2
	SMR	321	194	400	242	594	223
Private							
Not family in multi-person household	Obs	52	77	110	178	359	655
	Exp**	41.5	65.3	96.6	150.1	362.8	579.4
	SMR	125	118	114	119	99	113
Others	Obs	209	336	578	1082	3175	5963
	Exp**	200.1	365.4	634.5	1167.6	3434.4	6123.0
	SMR	104	92	91	93	92	97

Women													
Non-private Homes	Obs	67		90		139		137		105		134	
	SMR		180		181		196		164		164		171
	Exp**	37.3		49.8		70.8		83.6		64.2		78.3	
Resident in hospitals	Obs	42		34		71		59		139		121	
	SMR		316		234		302		267		301		220
	Exp	13.3		14.5		23.5		22.1		46.2		55.0	
Visitors in hospitals	Obs	18		10		36		11		190		110	
	SMR		400		145		878		224		597		274
	Exp**	4.5		6.9		4.1		4.9		31.8		40.1	
Private not family in multi-person household	Obs	58		128		170		279		636		1142	
	SMR		94		121		106		102		96		100
	Exp**	61.5		106.0		160.6		274.3		659.5		1146.4	
Other	Obs	128		290		459		929		2459		5141	
	SMR		80		91		89		93		86		95
	Exp**	160.6		317.9		513.2		994.6		2849.3		5431.8	

* Migration within England and Wales in the five years before census.

** In this table, expected deaths are calculated separately for each sex using death rates (in five-year age groups) for all men or women in the Longitudinal Study.

Table 4.5 Mortality of male five-year migrants by area of origin and destination

Grouped region of residence (1966)			Migrants between standard regions								Migrants within standard regions
			Grouped region of residence (1971)								
			South and east		Central		North and west		All in England and Wales		
South and east	Obs	SMR	104	79	37	98	31	103	172	87	1214	95
	Exp		131.0		37.6		30.2		198.8		1284.3	
Central	Obs	SMR	40	81	2	86	29	99	91	87	697	109
	Exp		49.5		25.6		29.3		104.4		636.7	
North and west	Obs	SMR	30	83	24	118	20	111	74	99	749	120
	Exp		36.2		20.3		18.0		74.5		624.6	
All in England and Wales	Obs	SMR	174	80	83	99	80	103	337	89	2660	104
	Exp		216.7		83.5		77.6		377.6		2545.6	

between grouped regions. The table shows that the gradient between the south and east and the north and west is greater by region of residence in 1971 than by region of residence in 1966. For men the SMRs were 87 and 99 based on residence in 1966 and 80 and 103 based on residence in 1971; for women there were 80 and 108 based on 1966 and 79 and 119 based on 1971.

The above discussion of selection and migration has shown the importance of selection to the mortality of migrants, and Table 4.5 confirms previous suspicions that migrants to the south and east tend to be healthier than migrants away from the south and east and migrants leaving the north and west have lower mortality than those coming to the north and west. Nevertheless, we have yet to measure the impact of migration on regional mortality patterns. This is achieved in Table 4.6, which compares the mortality of all people in the Longitudinal Study—including people whose address was

Table 4.6 SMRs of people of all ages by region of residence in 1966 and by region of residence in 1971

Region	Men		Women	
	1966 Residents	1971 Residents	1966 Residents	1971 Residents
South-west	93	93	95	95
South-east	91	91	93	93
East Anglia	95	94	84	81
West Midlands	103	104	103	103
East Midlands	102	102	101	101
Yorks and Humberside	103	103	110	109
North	109	110	107	107
North-west	114	114	110	110
Wales	116	115	111	113
All England and Wales	100	100	100	100

constant for the five years before the census and those who moved within regions—by region of residence in 1966 and region of residence in 1971. The table shows that only for women in East Anglia is there a change in SMR of more than two points. So that while we have been able to show that selection effects do exist, we must conclude that because of the low level of inter-regional migration they are not important to regional gradients in mortality. It remains to be seen whether they can be shown to be of greater significance in the study of smaller geographic areas, such as new towns, retirement towns, or inner cities.

Selection and marriage

Differences in mortality by marital condition have also been recognized for well over 100 years. Nevertheless, it is not clear from publications if these differences reflect Farr's (1859) suggestion that: 'marriage is a healthy state. The single individual is more likely to be wrecked on his voyage than the lives joined together in matrimony.' Berkson (1962), remarking on the similarities of differentials for different causes of death, suggested that errors in the method may underline the differentials. On the other hand, Spiegelman (1960) suggested that: 'the lower mortality of the married man may be due to a selection process on which those in poor health remain single. Furthermore, the married may receive better care to protect health or in case of illness. These are environmental advantages lost to the widowed and divorced.'

Such comments are usually based on traditional mortality data that do not permit analysis by 'time since change in status' or by 'duration of status'. The first of these was considered in the classic study of Young *et al*. (1963), which identified a sample of men at the time they became widowered and followed them up to death. This study highlighted what became known as the 'bereavement effect', which described high mortality shortly after the death of the spouse. It is not clear whether this short-term rise in mortality reflected the mortality of a group of people who were sick before their spouse died and whose condition deteriorated once their source of support was removed, or whether it simply reflected the direct effects of depression after the loss of a spouse. Nevertheless, it is the only effect that has been related in time to marital condition. And, since the effect only affects mortality within six months of the death of a spouse, it can be seen that it does not explain the high mortality of widows.

Mortality differentials by marital condition obtained from the OPCS Longitudinal Study are similar, but not as wide as, those obtained using more traditional methods. The main difference between results is that the Longitudinal Study indicates only a very narrow gradient for women aged over 45. Over 65, single women fare as well as married women.

Table 4.7 describes the change in gradient over the period of follow-up for the first five years of the Longitudinal Study. Only for widowered men is there a clear indication that the SMR is changing with duration of follow-up; this change is most pronounced in those aged 65–74 but it is also evident for those aged 75 and over. The census design of the study means that this group of widowered men describes the nett effect of two selection processes—the selection-in of people who are becoming widowed and the selection-out of those who are remarrying. The nett effect gives the appearance of ill-health selection predominating. The SMRs for married men aged 15–44 and for those aged 75 and over also provide early signs of selection—although not as pronounced as those for widowered men. The change in SMRs at older ages

Table 4.7 Mortality by sex, age, marital status, and year of death

Marital status	Men				Women			
	15–44	45–64	65–74	≥75	15–44	45–64	65–74	≥75
Single								
1971–2	132	134	111	104	131	102	90	99
1973	117	135	121	110	132	101	110	93
1974–5	119	136	117	107	114	105	93	97
Married								
1971–2	84	95	96	92	85	96	101	100
1973	87	96	96	95	88	98	93	93
1974–5	89	95	99	98	96	97	100	97
Widowed								
1971–2	*	146	122	110	*	111	103	101
1973	*	110	118	108	*	107	106	104
1974–5	*	134	92	104	*	112	104	102

* Only three men and two women died in 1971–5.

may reflect the impact of relatively healthy men remarrying after their spouse died. At younger ages the selection is probably that pointed to by Spiegelman (1960) of those in poor health remaining single.

The table also indicates a selection effect operating for married women aged 15–44 but not in other age groups, nor for widowed and single women. The difference between men and women who are widowed is not surprising because the proportion of men at older ages who are widowered is much lower than that for women. The sex differential in death rates leads to most husbands dying before their wives, and, because there are many more women than men alive at older ages, the remarriage rate for widowers is much higher than that for widows.

The 1971 census also asked women under 60 several questions about their marriage and fertility history. These questions in principle permit the analysis of death rates by date of first marriage termination, thereby enabling us to look for effects similar to the bereavement effect noted earlier. In practice, however, the death rates are low for those ages, and so the analysis is based on too few deaths for meaningful assessments to be made (Table 4.8).

An alternative approach that should prove more fruitful will be an analysis of survivorship after the death of a spouse in the mould of the study of Young *et al.* (1963) mentioned earlier. These data will shortly be available in the longitudinal study and will be used to see whether the bereavement effect is found for women whose husbands have died and also to see how it is related to the person's age when the spouse died.

Table 4.8 Mortality of women aged under 60
at census, by year of termination of
first marriage

Marital status (1971)	Year of first marriage termination	Deaths		SMR
		Obs	Exp	
Widowed	1971 (Jan—April)	6	5.2	115
	1970	17	15.3	111
	1969	14	13.6	103
	Pre-1969	136	119.2	114
	Not stated	23	15.6	147
Divorced	1971 (Jan–April)	1	1.3	77
	1970	3	2.8	107
	1969	2	2.0	100
	Pre-1969	46	31.7	145
	Not stated	9	2.8	321
Married second or subsequent marriage	1971 (Jan–April)	—	0.4	—
	1970	3	1.1	273
	1969	3	1.6	188
	Pre-1969	95	102.2	93

Marital status and selection into a non-private household

In the earlier discussion we focused on people who were enumerated in non-private households and separated them by type of institution and length of time in the institution to identify groups that included many people with acute or chronic sickness. Figure 4.4 contrasts the mortality of the single, married, and widowed who were enumerated in non-private households in 1971. It is not surprising that each group shows the time trends expected of people selected on the basis of ill health. The figure is of interest because it shows how by studying the changes in SMR over time, inferences may be drawn about the underlying selection processes. In each quadrant of the diagram the married have appreciably higher mortality than the single. The decline with time for the married and, to a lesser degree, the widowed, is similar to the decline observed in Fig. 4.1 for the acutely sick. The decline with time for the single is, however, more in line with that expected for the chronically sick.

Analysis of these data by type of institution confirms that a higher proportion of the married than the widowed or the single in non-private households were visiting hospitals. Nevertheless, there is also clear evidence that the single in homes or resident in hospitals have lower levels of disability (as measured by subsequent mortality) than the widowed or the married in the corresponding type of institution. This difference probably reflects differences in support available to people of different marital status and shows how similar selection processes may operate to different degrees according to social and household circumstances. Single people with the

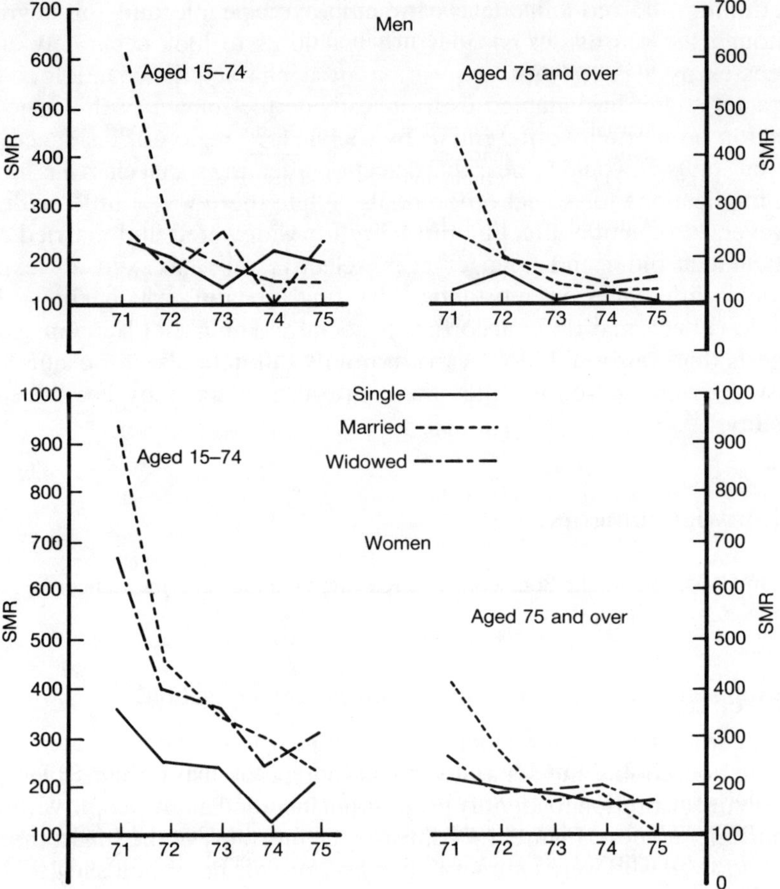

Fig. 4.4 Mortality in non-private households by sex, marital status, and year of death.

need for minimal support may not be able to find that support from their immediate environment when married and widowed people with similar needs can call on relatives for support.

Conclusions

Our definition of health-related mobility carries with it implications for the relationship between mortality and duration of follow-up. We have used the Longitudinal Study to see whether these implications are confirmed for characteristics that for many years have been associated with selection. These analyses confirm that health-related mobility is an important factor in determining the mortality levels for employed people and for migrants. Although the census design does not enable us to look separately at the processes associated with acquisition of a characteristic and loss of a characteristic, it has enabled us to investigate the impact of this aspect of selection on mortality differentials by social class, region of residence, and marital status. It would appear that selection affecting social class has important implications for social differentials. While there was a little evidence, however, of selection affecting mortality for widowered and married men, particular at older and younger ages, selective migration did not explain regional differences in mortality. The final section, which shows how selection effects may differ according to social and household circumstances, suggests that we should look at the mortality differentials—for example, by housing tenure to see whether these are also affected by health-related mobility.

Acknowledgements

AJF and AMA thank the Social Science Research Council and the Cancer Research Campaign for their support.

References

Berkson, J. (1962). Mortality and marital status—reflections on the derivation of etiology from statistics. *Am J. Public Health* **52**, 1318.

Bradford-Hill, A. (1925). *Internal migration and its effect upon the death rate: with special reference to the county of Essex*. MRC special report No. 95. HMSO, London.

Farr, W. (1859). The influence of marriage on the mortality of the French people. *Transactions of the National Association for the Promotion of Social Science*, 1858: **504**. Published in full in: *Influence of marriage on the mortality of the French people*. Savill and Edwards, London.

Fox, A. J. (1980). Prospects of measuring changes in differential mortality. In: *Proceedings of the meeting on socio-economic determinants and consequences of mortality in Mexico City, June 1979*. World Health Organisation/United Nations Organisation.

—— and Collier, P. F. (1976). Low mortality rates in industrial cohort studies due to selection for work and survival in the industry. *Br. J. Prev. Soc. Med.* **30**, 225–30.

—— and Goldblatt, P. O. (1982). *Socio-demographic differentials in mortality*. Longitudinal study series LS No. 1. HMSO, London.

Goldblatt, P. O. and Fox, A. J. (1979). Household mortality from the OPCS Longitudinal Study. *Population Trends* **14**, 21–7.

Goldsmith, J. R. (1975). What do we expect from an occupational cohort? *JOM* **17**, 126.

Office of Population Censuses and Surveys (1978). *Occupational mortality 1970–72. Registrar General's decennial supplement*. Series DS No. I. HMSO, London.

OGLE, W. (1885). Letter to the Registrar General on the mortality in the registration districts of England and Wales during ten years 1871–80. *Supplement to the 45th Annual Report of the Registrar General of Births, deaths and Marriages, in England* xxiii.

SPIEGELMAN, M. (1960). Factors in human mortality. In: *The biology of aging.* Symposium No. 6, 292. American Institute of Biological Sciences, Washington DC.

STEVENSON, T. H. C. (1913). *74th annual report of the Registrar General for the year 1911* lii. HMSO, London.

WELTON, T. A. (1872). The effects of migration in disturbing local rates of mortality as exemplified in the statistics of London and the surrounding country for the years 1851–60. *Journal of the Institute of Actuaries* **16** .

YOUNG, M., BENJAMIN, B., AND WALLIS, C. (1963). The mortality of widowers. *Lancet* ii , 454–6.

5

A study of the mortality of female asbestos workers

M. L. NEWHOUSE, G. BERRY, J. C. WAGNER, AND M. E. TUROK

A cohort study of over 900 women employed at an asbestos factory making both textiles and insulation materials is described. It extends the information about asbestos-related disease at this factory which was previously available only for male workers. The cohort was defined as all the women who started employment at the factory between 1936 and 1942 and the main analysis was of mortality up to the end of 1968. This analysis was made in relation to job, length of exposure, and age at first exposure. Compared with national rates there was excess overall mortality among those who worked in jobs with low to moderate exposure partly accounted for by deaths from cancer. In the group with severe exposure, who had worked in the factory for less than two years, there was an excess of cancer of the lung and pleura. However, the most marked increased mortality was in those with severe exposure who had worked for more than two years in the factory; in this group there were excess deaths from cancer of the lung and pleura, from other cancers, and from respiratory diseases. There were no significant trends of excess mortality with age at first exposure. The smoking habits of some of the deceased women were obtained and the indications were that the proportion of smokers in the cohort was higher than the national rate. This could account for some of the excess mortality but the trend of this excess with exposure indicated the role of asbestos. Necropsy reports and/or histological material were obtained for 43 per cent of those who had died. Three deaths registered as cancer of the pleura were identified as pleural mesothelial tumours; in all there were 11 mesotheliomas, six of pleural and five of peritoneal origin.

The mortality of male asbestos workers in mines and factories has been extensively investigated (Doll 1955; Selikoff, Churg, and Hammond 1964; McDonald *et al*. 1971). There is less opportunity to examine the mortality of female workers although there is considerable evidence that women exposed to asbestos are at risk of developing asbestos-related cancers and asbestosis. Keal (1960) reported a high incidence of peritoneal tumours among women patients suffering from asbestosis. Enticknap and Smither (1964) reported three female deaths in their series of 11 patients dying from peritoneal mesothelioma. Wagner, Sleggs, and Marchand (1960) found nine women with mesothelial tumours among 33 subjects in South Africa. Among 83 patients with a confirmed diagnosis of mesothelial tumour from the London

Hospital (Newhouse and Thompson 1965), 42 were females. In Canada, approximately one-third of the mesothelial tumours reported over a 10-year period occurred in women (McDonald *et al*. 1970). Mancuso and El-Attar (1967) included 228 women asbestos workers in their cohort followed for 25 years; eight of the 31 deaths that occurred during the period were due to asbestosis, six to pulmonary malignancies, and one to a peritoneal mesothelioma.

Definition of the cohort

Here we describe the mortality experience of women who have been employed in an asbestos factory. Methods of tracing the past employees of this factory have already been described (Newhouse and Williams 1967). The factory was in production for 56 years and employed more men than women. Reliable information was obtained about a high proportion of the men, but the methods failed to trace more than 30 per cent of the women employees. The principal difficulty was the change of name on marriage, subsequent to leaving the firm's employment. A further attempt has been made to trace women who started work at the factory during the seven years between 1 January 1936 and 31 December 1942, a period when the information available to ourselves was expected to link up with the information at the National Health Service Central Register (NHSCR). A pilot study of 100 of the women yielded information on 76 subjects. With this encouraging result the names, dates of birth, and addresses during employment of all 928 women who started work at the factory during this period were sent to the NHSCR. Six hundred and thirty-six (67 per cent) members of this cohort were identified as currently alive or, if deceased, a copy of the entry to the Register of Deaths was given. Forty-six (5 per cent) had either emigrated or joined the Armed Forces before 1952 and could not be followed further. A search of the marriage indices at Somerset House identified the married names of a further 63 women whose current status could then be determined by the NHSCR. An attempt was made to obtain information about past workmates from women who had been employed for long periods at the factory. Of the 64 women written to, only 21 replied and four were visited. Information obtained through them and a final search through factory records led to another 17 women being traced. At 31 December 1968, 576 of the 928 women were identified as alive and 140 dead. Thus 716 (77 per cent) had been traced. The dates of birth of six of the untraced women were not known and these women were excluded from the cohort at this stage.

Occupational exposure

One hundred and twenty-six of the 922 women in the cohort had been employed only in jobs with low to moderate exposure to asbestos dust. They were those employed in the office, canteen, laboratory, and print stores. A few with low exposure were in production departments, such as brake lining, where chrysotile asbestos was used, and rubber jointings, where crocidolite and chrysotile asbestos were used. The remainder had all worked in production departments, over 400 in the textile departments in carding, spinning, doubling, or weaving; 100 were in a mattress-making section. These are jobs where there is a statutory requirement under the Asbestos Industry Regulations (1931) for periodic medical examination. Crocidolite was used in the spinning and doubling department and chrysotile and amosite for carding, weaving, and mattress making. Two hundred and twenty-five other women were employed on sectional pipe making, preparing filters for gas masks, and in the yarn store. In these jobs also either crocidolite alone or both crocidolite and amosite were used. Methods of dust suppression were required in all areas, but women in the latter group of jobs were not required to have medical examinations. Workers in both groups of job were considered to be severely exposed to dust. The textile and mattress-making departments at this factory closed in the mid 1950s.

Method of analysis

The results have been assessed by comparing the number of observed deaths with the number of expected deaths. The 'expected' deaths were calculated by the 'man-years' method (Case and Lea 1955), multiplying years of risk by death rates.

Causes of death, as recorded on the death certificate, were coded according to the Seventh Revision of the International Classification of Diseases (World Health Organization 1957) and the groups of causes considered were: all cancers (ICD 140–205), cancer of the lung and pleura (ICD 162, 163), cancer of the gastrointestinal tract (ICD 150–157), cancer of the ovary and fallopian tubes (ICD 175), and respiratory diseases excluding lung cancer (ICD 001–008, 240, 241, 465, 490–493, 500–502, 518–527).

Death rates for England and Wales were taken from the tables compiled for studies of this kind by Case and his colleagues (General Register Office 1957; Case and Harley 1958; Case, Coghill, and Harley 1968a, b). These tables give age-specific death rates with age in five-year groups up to 85, and calendar year also in five-year groups up to 1961 and 1965. The period of study extended beyond this and for 1966 to 1968 the rates for 1961 to 1965 were used.

Each woman was considered at risk from the date of first employment at the factory until 31 December 1968 for those traced as alive, the date of death for those known to be dead, or the date of leaving employment at the factory for those untraced.

Excess mortality has been tested by treating the observed number of deaths as a Poisson variable with expectation equal to the man-years expected number of deaths. Although it can be shown that this test may lead to the significance of excess mortality being exaggerated, the size of such a bias is unlikely to be of practical importance.

Observed and expected mortality

For the analysis, the women were grouped into those with periods of employment lasting up to two years, and those with longer exposure. Each of these groups was further subdivided into two categories, those with low to moderate exposure and those with severe exposure to asbestos dust. The proportion of untraced women varied. It was lowest among women with low to moderate exposure and periods of employment of two years or longer, and highest in the group with severe exposure but shorter periods of employment at the factory (Table 5.1). A third factor examined was age at start of exposure grouped as 15–24, 25–34, and 35+.

In the total cohort there was no significant excess of deaths with gastro-intestinal cancer (observed 11, expected 7.3, $P = 0.12$) or cancer of the ovary (observed 4, expected 2.1, $P = 0.16$) and these two causes have not been detailed in the following mortality tables, although they are discussed where relevant information was obtained.

Table 5.1 Number of women in cohort and number untraced

Length of employment (yr.)	Degree of exposure			
	Low to moderate		Severe	
	Total no.	No. untraced	Total no.	No. untraced
Less than 2	83	17 (20%)	557	153 (27%)
More than 2	43	2 (5%)	239	34 (14%)

Low to moderate exposure

Among this group of women there was little difference in the mortality experience of those with short and those with long exposure and, as the group comprised only 126 individuals in all, the results have been combined (Table 5.2). There was a general excess of deaths, which was partly accounted for by deaths from cancer—two from cancer of the lung and pleura and eight from other cancers; this latter group included three gastro-intestinal cancers. There was also an excess of deaths from diseases other than cancer or respiratory disease, but this was not significant and did not suggest that any particular cause of death was implicated. The recorded

Table 5.2 Mortality of women with low to moderate exposure

Registered cause of death	All periods of employment (126 women)	
	Obs.	Exp.
All causes	29*	18.1
Cancer of lung and pleura	2*	0.3
Other cancer	8	4.4
Respiratory disease excluding cancer	2	2.2
Other diseases	17	11.2

* $p < 0.05$.

occupations of the two women who died from cancer of the lung were office and canteen worker, jobs which do not suggest an occupational exposure, but nothing is known of possible contact with other workers or exposure in the home.

Severe exposure

This group is larger and is, therefore, divided into those with up to two years' employment and those with longer periods (Table 5.3). While there was a highly significant excess of deaths from cancer of the lung and pleura in those with the shorter periods of employment, the most marked effect was in those with longer periods of employment. Here there was a significant excess of deaths from cancer of the lung and pleura, other cancers, and respiratory diseases. Among the six dying from cancer of the lung and pleura in the group

Table 5.3 Mortality of women with severe exposure

Registered cause of death	Duration of employment			
	Less than 2 yr (557 women)		More than 2 yr (239 women)	
	Obs.	Exp.	Obs.	Exp.
All causes	55	49.9	56***	24.5
Cancer of lung and pleura	6***	1.0	14***	0.5
Other cancer	16	12.4	17***	6.1
Respiratory disease excluding cancer	10	7.4	11**	3.6
Other diseases	23	29.1	14	14.3

** $p < 0.01$
*** $p < 0.001$.

with short periods of employment were three who were proved to have pleural mesothelioma, one of whom had worked in the factory for only six weeks. The 16 other cancers included five gastrointestinal and one cancer of the ovary. In the group with longer employment, the 17 deaths from other cancers included two registered as dying from peritoneal mesotheliomata, three from gastrointestinal cancer, three from cancer of the ovary, and three registered as carcinomatosis. In one of the death entries giving carcinomatosis as the primary cause of death, the underlying cause of death was given as 'adenocarcinoma probably bronchus' and in the other two, carcinoma of the ovary was suggested as the primary site of the cancer. In the group of 11 certified as dying from respiratory diseases, the primary cause of death was certified in three as asbestosis.

Excess annual death rate

In order to obtain a measure of the excess mortality in absolute terms, excess annual death rates were calculated. These rates are obtained by dividing the excess number of deaths (observed minus expected) for any cause, by the number of subject-years at risk in the group. For lung cancer, for the group with low to moderate exposure, the mean value of the excess annual death rate was 63 per 100 000 years' exposure, for the severely exposed group with short exposure 44, and for the group with severe exposure for two years or longer 241. The differences between the three groups were significant ($P < 0.01$). The third group had a significantly higher excess rate ($P < 0.001$) than the other two.

Table 5.4 Mortality by age at which employment at factory was commenced

Registered cause of death	Age started employment					
	15–24 (516 women)		25–34 (210 women)		35+ (196 women)	
	Obs.	Exp.	Obs.	Exp.	Obs.	Exp.
All causes	45***	22.2	25	17.7	70*	52.5
Cancer of lung and pleura	12***	0.4	4**	0.5	6***	0.9
Other cancer	15***	5.1	9	5.4	17	12.3
Respiratory disease excluding cancer	10*	5.4	4	2.4	9	5.4
Other diseases	8	11.3	8	9.4	38	33.9

  * $p < 0.05$
 ** $p < 0.01$
*** $p < 0.001$

Effect of age at first exposure

The mortality was also examined in relation to the age of the women on first employment in the factory (Table 5.4). Fifty-six per cent started work before the age of 25 and two-thirds of these were under 20; a further 23 per cent were first employed between the ages of 25 and 34; and 21 per cent were 35 or older when first engaged. All degrees and lengths of exposure have been combined in Table 5.4. In the three age groups the proportions of women in the group with greatest excess mortality, that is those with severe and long exposure, were 27, 21, and 30 per cent. After adjusting for degree and length of exposure, the mortality pattern is not very different from that given in Table 5.4 and, hence, conclusions may safely be based on that table. Although the most significant excess mortality occurred in the youngest group, this was due to the larger size of this group and to the lower natural death rates of the younger women. In absolute terms, there are no significant differences in the excess mortality of the three age groups; for lung cancer the excess annual rates were 108, 76, and 122 per 100 000 years at risk.

Length of follow-up

The distribution of observed and expected deaths was also calculated according to the length of follow-up since first exposure to asbestos in the factory (Table 5.5). The maximum length of follow-up was 33 years. There was no excess mortality among deaths occurring within 10 years of first exposure. For a follow-up period of between 10 and 20 years, the excess lung cancer mortality was beginning to show, while after 20 years the mortality was much more marked. This is in agreement with the mortality analysis of

Table 5.5 Mortality by length of follow-up since first exposure

Registered cause of death	Years since first exposure					
	Less than 10 (922 women)		10 to 20 (692 women)		More than 20 (655 women)	
	Obs.	Exp.	Obs.	Exp.	Obs.	Exp.
All causes	25	23.7	36	29.3	79***	39.4
Cancer of lung and pleura	0	0.2	3*	0.6	19***	1.1
Other cancer	5	4.2	10	7.8	26***	10.8
Respiratory disease excluding cancer	9	6.1	5	3.4	9*	3.7
Other diseases	11	13.2	18	17.5	25	23.8

* $p < 0.05$.
*** $p < 0.001$.

male asbestos workers (Newhouse 1969) in which excess mortality was not observed until more than 16 years after first exposure.

Smoking habits

An attempt was made to ascertain the smoking habits of the women who had died. Little information could be obtained about those who died before 1960, but, from hospital and ward notes, general practitioners, and relatives of the deceased, it was possible to ascertain the smoking history of 93 out of 103 women who died between 1960 and 1970. This particular study, therefore, excludes the earlier deaths but includes deaths notified as occurring in 1969 and 1970. Out of the 93 deceased women, 60 (65 per cent) were reported to have been smokers. Taking account of the age distribution of these women and using the smoking prevalences for 1956 for women in the United Kingdom given by Todd (1969), 39 (42 per cent) of these women would have been expected to be smokers. However, it is well established that smoking results in a shortening of life (Royal College of Physicians 1971) so that it would be expected that the proportion of smokers in the above group of deaths would be higher than in the whole cohort. Whether it would have been as high as 65 per cent if the smoking habits of the cohort were similar to a national rate requires knowledge of the age-specific death-rates for non-smokers and smokers and, as these data are sparse for women, no attempt has been made to examine this point further. It is probable that in this cohort of women there was a higher proportion of smokers than in the whole female population of the United Kingdom.

There were 22 women in the 1960–70 group certified as dying of cancer of the lung and pleura and, of these, six were proved to have a pleural meso-thelioma. The smoking habits of these 22 women were four non-smokers, four light smokers (1–9 cigarettes/day), 10 moderate smokers (10–19 cigarettes/day), and four heavy smokers (20+ cigarettes/day). Only two of the women who died of pleural mesothelioma were non-smokers.

Validation of the certified cause of death

In this series, as in the series of deaths of male asbestos workers (Newhouse and Wagner 1969), an attempt was made to obtain information additional to the certified cause of death in order to validate the suggested diagnosis. Hospitals and coroners were approached and asked for copies of post-mortem reports, and histological specimens were requested from the pathologists performing the necropsy. In the case of women who had died at home and where the death had been certified as due to cancer of any site, the general practitioner was asked for the name of the hospital where the patient had been treated and a request was made for histological material from biopsy or surgical specimens. There were 140 deaths in the series. In many, particularly where death had occurred before 1950, records and specimens had not been preserved, but necropsy reports were obtained in 51 (Table 5.6) and necropsy records together with histological material in 30 of these; biopsy material was obtained in a further nine of the series. The London Hospital and the London Pneumoconiosis Panel together contributed one-third of the necropsy reports and histological material. Eleven necropsy reports came from coroners and the remaining material from 20 other hospitals. All histological material was reviewed by one of us (J.C.W.) and all suspected mesothelial tumours were referred to the members of the mesothelioma panel. The opinion of Dr F. A. Langley of Manchester University was obtained on tumours of possible ovarian origin.

Additional information was received about all but four of the group of lung cancers (Table 5.6). Three of the 22 deaths originally certified as cancer of the lung or pleura were reclassified as pleural mesothelioma and this diagnosis was confirmed in a further three cases. There were five deaths certified as peritoneal mesothelioma. This diagnosis was confirmed in each case and no additional tumours were identified.

We were not successful in obtaining any necropsy reports for the 11 women who died of gastrointestinal cancer. Biopsy material was obtained from four; in two, this was adequate to confirm the certified cause of death given as carcinoma of the oesophagus; in the other two, the specimens were too poorly preserved to reveal the tumour type. There were four deaths registered as cancer of the ovary; in the two cases where histological material was available, the registered cause of death was confirmed. Further information was received for only two of the six deaths certified as carcinomatosis. In both cases, examination of the protocols and histology suggested an ovarian origin of the tumour.

Sections of the lung tissue were examined in 23 of the series; evidence of asbestosis was found in all but two. The cases for which lung tissue was available for examination included six of those certified as dying of cancer of the lung, and four of pleural mesothelioma, one of peritoneal mesothelioma, and two of asbestosis. In the remaining 10 sections, the certified cause of death was cancer of the breast (2), cerebral tumour (1), uraemia (1), and respiratory conditions (6).

Discussion

The mortality of the women workers at this factory appears to follow a pattern similar to that of the male workers (Newhouse 1969). In both groups there is a highly significant excess of cancers of the lung and pleura. Among workers followed for more than 20 years, about 1 per cent of the men and 1.5 per cent of the women died of mesothelial tumours. However, in this series, only 77 per cent of the members of the cohort were traced. If the 206 untraced women had experienced a similar mortality to the traced, then the expected number of deaths among them would be about 29. If the true

Table 5.6 Number of necropsy and pathological specimens received

Diagnostic category	Necropsy report and histology	Histology only	Necropsy report only	No further information	Total
Cancer of lung and pleura (ICD 162, 163)	16	1	1	4	22
Peritoneal mesothelioma (ICD 158)	3	1	1	0	5
Gastrointestinal cancer (ICD 150–157)	0	4	0	7	11
Carcinomatosis (ICD 199)	1	1	0	4	6
Ovarian tumour (ICD 175)	1	1	0	2	4
Cancer not otherwise specified	2	1	3	9	15
Respiratory disease excluding cancer	4	0	2	17	23
Other non-malignant diseases	3	0	14	37	54
Total	30	9	21	80	140

number of deaths was very much different from this, then the incompleteness of follow-up would have introduced a bias. As an extreme example, suppose none of the untraced women had died, then the expected numbers of deaths would be approximately: total deaths 113 (observed 140), all cancers 30 (observed 63), lung cancer 2.2 (observed 22). Hence, even in this highly unlikely event, there would still have been excess mortality. The method of tracing was such that the fact that a woman had died did not materially increase the likelihood of her being traced.

The use of rates for 1961–5 for calculating expected deaths for 1966–8 may be criticized, but examination of age-specific death rates for 1962, 1963, and 1967 (General Register Office 1964–70) showed the error in the calculated number of expected deaths to be about 1 per cent or less for total deaths and all separate causes except for lung cancer. Here there was an increase of 25 per cent in the rates between 1963 and 1967, which modified the expected number of deaths by 5 per cent; however, this results in the negligible change from 1.8 to 1.9 in the expected figure. Secondly, the uses of rates from a smaller and geographically similar area, such as Greater London, might have been more appropriate than the national rates. The SMR for lung cancer for Greater London was 137 per cent in the quinquennia 1963–7 (General Registrar Office 1964–70). If this figure had been applied to the expected number (1.8), we would have expected 2.4 deaths. This would still have been a highly significant excess.

Eleven mesothelial tumours were identified, six pleural and five peritoneal. Six of these tumours occurred in women who had worked for less than two years in the factory. The proportion of peritoneal tumours is high compared with several series of necropsy cases. For example, McEwen *et al.* (1970), in a Scottish study of a predominantly male population, found only three cases arising from the peritoneum out of 80 mesotheliomas; Ashcroft and Heppleston (1970) found three out of 23 in a Tyneside study; and Thomson (1970) found three out of 17 in Cape Town. However, higher proportions of peritoneal tumours have been found in cohort studies of asbestos workers. In the study of males from the same factory as the female workers, 13 peritoneal mesotheliomas out of a total of 19 were reported (Newhouse and Wagner 1969) and more recent figures for this study are 30 out of 49. Selikoff, Hammond, and Churg (1970) found 16 peritoneal mesotheliomas out of a total of 22 mesotheliomas occurring in their cohort of male asbestos insulation workers in the New York area, and these authors suggested that the differences between the two types of study could be the result of the greater exposure in the two cohort studies, pleural mesotheliomas being associated with lesser exposure and peritoneal with greater.

There were four deaths in all registered as cancer of the ovary; three of these occurred among the women with severe and long exposure and, compared with the expected number of 0.6 in this particular group, this was a significant finding ($P = 0.0025$). The histological review suggested that at least two of the other deaths in the group which were registered as carcinomatosis were due to this cause. These figures are too small for definite conclusions but, in view of the experimental work (Graham and Graham 1967), which suggested that ovarian tumours can be produced in laboratory animals after the intraperitoneal injection of tremolite asbestos, collection of further data is of importance.

There was a highly significant excess of deaths from cancer of the lung, occurring chiefly in those with severe exposure to asbestos dust. As in the study of male asbestos workers (Newhouse 1969), it was found in both the group with short and the group with long exposure, but the excess lung cancer risk which was found in the low-moderate exposure group was not observed in the male study. This difference could be due to the suggested higher proportion of smokers in the cohort than in the general female population. Our evidence on smoking is incomplete and obtained at second hand, but Todd (1966) found that recalls of smoking habits by relatives were a reasonably accurate substitute for recalls by the smokers themselves. The much higher excess lung cancer risk in those women with severe exposure for a shorter period and the occurrence of pleural mesotheliomas confirms that the exposure to asbestos was responsible for an appreciable proportion of the excess lung cancers. Comparisons within the severe exposure group also show that asbestos exposure resulted in excess deaths from other cancers and deaths with respiratory diseases.

The excess of lung cancer deaths is much more striking in the women than in the men (Newhouse 1969). In the severely exposed group, the ratio of observed to expected lung cancer is 13 for the women but only 2.5 for the men. Jacob and Anspach (1965) found a similar result with ratios of 16 and 2.1 for women and men respectively. However, if one considers the absolute excesses, the rates are similar for men and women and it is therefore suggested that exposure to asbestos produces a similar effect in men and women as far as lung cancer is concerned, but that the effect is more easily detected in a cohort of women because of the lower lung cancer rates in the general population.

The presentation in Table 5.4 of the effect of age at first exposure may be criticized in that no allowance was made for the length of follow-up, which we have shown to be important. We would expect that the average length of follow-up would decrease with age at first exposure because of the higher mortality of older women and that Table 5.4 would therefore underestimate an age effect. As another way of looking at this aspect, length of follow-up has been allowed for by apportioning the excess number of lung cancers for each length of follow-up between the three age groups according to the number of subject years. For the 15–24 year group this gave an expected excess of 12.2, for the 25–34 year group of 4.6, and for the 35+ group of 3.4. The observed excesses were 11.6, 3.5, and 5.1, which are 95, 76, and 150 per cent of the expected figures. Therefore, in the oldest group there is some support of an age effect on mortality from lung cancer. However, the trend, in contrast to the findings of Knox *et al.* (1968), is not at all significant. Evidence on the effect of age at first exposure has recently been reviewed by Doll (1970) and found to be conflicting.

Selikoff, Hammond, and Churg (1968) found that male asbestos workers with a history of regular cigarette smoking had about 90 times the risk of dying of bronchogenic carcinoma as men who neither smoked cigarettes nor were occupationally exposed to asbestos dust. More information is needed on the combined effect of smoking and asbestos exposure on mortality and our study of both male and female workers is being extended in co-operation with the Department of Health and Social Security by carrying out a survey of the smoking habits of the surviving individuals in both cohorts. When this is complete, it may be possible to estimate how much of the excess risk of dying from lung cancer is due to smoking and how much is due to asbestos exposure. The cohort study is also continuing, notification of deaths being received within a short period of their occurrence. More information, both about the effect of smoking in asbestos-exposed workers and about the occurrence of other possibly asbestos-related tumours, should be available within the next 10 years.

Acknowledgements

We are grateful to the late Dr W. J. Smither and the Cape Asbestos Company for their invaluable assistance and co-operation; to the Registrar-General and the National Health Service Central Registrar for supplying copies of the death entries and for tracing the subjects in the study; to Dr F. A. Langley for examining some of the pathological material; and to Dr J. C. Gilson and Professor R. S. F. Schilling for their advice and encouragement.

The study was undertaken with the assistance of a grant to M.L.N. from the Medical Research Council.

References

Asbestos Industry Regulations (1931). *Asbestos Industry Regulations: Statutory Rules and Orders* No. 1140, HMSO, London.

ASHCROFT, T. AND HEPPLESTON, A. G. (1970). Mesothelioma and asbestos on Tyneside—a pathological and social study. In *Pneumoconiosis. Proceedings of the International Conference, Johannesburg 1969* (ed. H. A. Shapiro) pp. 177–9. Oxford University Press, Cape Town.

CASE, R. A. M., COGHILL, C., AND HARLEY, J. L. (1968a). *Supplement to cancer death rates by site, age, and sex, England and Wales. Death rates for 1956–1960 and 1961–1965.* Chester Beatty Research Institute, London.

——, ——, AND —— (1968b). *Supplement to death rates by age and sex for tuberculosis and selected respiratory disease, England and Wales, 1911–1955. Death rates for 1956–1960 and 1961–1965.* Chester Beatty Research Institute, London.

—— AND HARLEY, J. L. (1958). *Death rates by age and sex for tuberculosis and selected respiratory diseases, England and Wales, 1911–1955*. Chester Beatty Research Institute, London.

—— AND LEA, A. J. (1955). Mustard gas poisoning, chronic bronchitis, and lung cancer. *Brit. J. prev. soc. Med.* **9**, 62–72.

DOLL, R. (1955). Mortality from lung cancer in asbestos workers. *Brit. J. industr. Med.* **12**, 81–6.

—— (1970). Cancer and ageing: the epidemiological evidence. *Dorn Memorial Lecture*. 10th International Conference of the International Union against Cancer, Houston, U.S.A.

ENTICKNAP, J. B. AND SMITHER, W. J. (1964). Peritoneal tumours in asbestosis. *Brit. J. industr. Med.* **21**, 20–31.

General Register Office (1957). *Studies on medical and population subjects, No. 13: Cancer statistics for England and Wales, 1901–1955* (by A. McKenzie, with additional material by R. A. M. Case and J. T. Pearson). HMSO, London.

—— (1964–70). *The Registrar General's statistical review of England and Wales for the years 1962–1968. Part I, tables, medical*. HMSO, London.

GRAHAM, J. AND GRAHAM, R. (1967). Ovarian cancer and asbestos. *Environ. Res.* **1**, 115–28.

JACOB, G. AND ANSPACH, M. (1965). Pulmonary neoplasia among Dresden asbestos workers. **132**, 536–48.

KEAL, E. E. (1960). Asbestosis and abdominal neoplasias. *Lancet* **2**, 1211–16.

KNOX, J. F., HOLMES, S., DOLL, R., AND HILL, I. D. (1968). Mortality from lung cancer and other causes among workers in an asbestos textile factory. *Brit. J. industr. Med.* **25**, 293–303.

MCDONALD, A. D., HARPER, A., EL ATTAR, O. A., AND MCDONALD, J. C. (1970). Epidemiology of primary malignant mesothelial tumors in Canada. *Cancer (Philad.)* **26**, 914–19.

MCDONALD, J. C., MCDONALD, A. D., GIBBS, G. W., SIEMIATYCKI, J., AND ROSSITER, C. E. (1971). Mortality in the chrysotile asbestos mines and mills of Quebec. *Arch. environm. Hlth* **22**, 677–86.

MCEWEN, J., FINLAYSON, A., MAIR, A., AND GIBSON, A. A. M. (1970). Mesothelioma in Scotland. *Brit. med. J.* **4**, 575–8.

MANCUSO, T. F. AND EL-ATTAR, A. A. (1967). Mortality pattern in a cohort of asbestos workers. A study based on employment experience. *J. occup. Med.* **9**, 147–62.

NEWHOUSE, M. L. (1969). A study of the mortality of workers in an asbestos factory. *Brit. J. industr. Med.* **26**, 294–301.

—— AND THOMPSON, H. (1965). Mesothelioma of pleura and peritoneum following exposure to asbestos in the London area. *Brit. J. industr. Med.* **22**, 261–9.

—— AND WAGNER, J. C. (1969). Validation of death certificates in asbestos workers. *Brit. J. industr. Med.* **26**, 302–7.

—— AND WILLIAMS, J. M. (1967). Techniques for tracing past employees; an example from an asbestos factory. *Brit. J. prev. soc. Med.* **21**, 35–9.

Royal College of Physicians (1971). *Smoking and Health Now* pp. 24–34. Pitman Medical and Scientific, London.

SELIKOFF, I. J., CHURG, J., AND HAMMOND, E. C. (1964). Asbestos exposure and neoplasia. *J. Amer. med. Ass.* **188**, 22–6.

——, Hammond, E. G., and Churg, J. (1968). Asbestos exposure, smoking, and neoplasia. *J. Amer. med. Ass.* **204**, 106–12.

——, ——, —— (1970). Mortality experience of asbestos insulation workers 1943–1968. In *Pneumoconiosis. Proceedings of the International Conference, Johannesburg, 1969* (ed. H. A. Shapiro) pp. 180–6. Oxford University Press, Cape Town.

Thomson, J. G. (1970). The pathological diagnosis of malignant mesothelioma of pleura and peritoneum. In *Pneumoconiosis. Proceedings of the International Conference, Johannesburg 1969* (ed. H. A. Shapiro) pp. 150–4. Oxford University Press, Cape Town.

Todd, G. F. (1966). *The Reliability of statements about smoking habits. Supplementary Report*. Res. paper No. 2a. Tobacco Research Council, London.

—— (1969). *Statistics of Smoking in the United Kingdom*. Res. paper No. 1 (5th edn) p. 26. Tobacco Research Council, London.

Wagner, J. C., Sleggs, C. A., and Marchand, P. (1960). Diffuse pleural mesothelioma and asbestos exposure inNorth Western Cape Province. *Brit. J. industr. Med.* **17**, 260–71.

World Health Organization (1957). *Manual of the International Statistical Classification of Diseases, Injuries, and Causes of Death*. WHO, Geneva.

6

Previous hospital care as a risk factor for pneumonia. Implications for immunization with pneumococcal vaccine

D. S. FEDSON AND THE LATE J. A. BALDWIN

In the Oxford Record Linkage Study population in 1970, seven hundred and ninety-three persons were hospitalized for or died as a result of pneumonia. Thirty-six per cent who survived and 49 percent who died had been discharged from hospital within the previous five years. For the period 1963 through 1970, cohort analysis determined the probability of subsequent readmission and/or death caused by pneumonia within the next five years for patients discharged with any condition and with high-risk conditions only. From this analysis, it was estimated that pneumococcal immunization of relatively few discharged patients would prevent each subsequent readmission and death from pneumococcal pneumonia. These results suggest that, in addition to age and underlying medical condition, patterns of previous hospital care can be used to identify many persons at increased risk of developing pneumonia. If current patterns of previous hospital care are similar to those found in Oxfordshire, physicians should consider giving pneumococcal vaccine to patients who are discharged from hospitals.

Pneumonia is responsible for substantial morbidity and mortality in the United Kingdom. The 1978 mortality figures for England and Wales indicate that 52 453 persons died with pneumonia, accounting for approximately 9 per cent of deaths from all causes (Office of Population Censuses and Surveys 1980). In almost 95 per cent of pneumonia deaths, the causative organism was not specified. Pneumococcal infections were reported for only 4.7 per cent of all pneumonia deaths. It is unknown what proportion of deaths each year due to unspecified organisms are caused by pneumococci, but it is reasonable to assume that the percentage is not insignificant.

Pneumococcal vaccine contains 14 capsular antigens representing serotypes responsible for approximately 80 per cent of bacteraemic pneumococcal infections in the United States (ACIP 1981). The same serotypes may be responsible for a similar proportion of serious pneumococcal infections in the United Kingdom (Turk 1978). Clinical trials in younger adults have demonstrated a vaccine efficacy of 80 per cent or more in preventing clinical illness from vaccine serotype organisms (Austrian, Douglas, Schiffman, *et al*. 1976). Other methods have estimated an efficacy of 60 per cent or less in patients who, for the most part, have undergone splenectomy or who have

Reprinted from *Journal of the American Medical Association* **248** (16), 1989–95 (1982).

conditions that severely compromise host-defence mechanisms (Broome, Facklam, and Fraser 1980). These lower estimates are of uncertain value for other persons who are immunocompetent but are at increased risk for severe pneumococcal infection because of age and/or other chronic medical conditions (Austrian 1980). In the absence of clinical trial data in these groups, there remains considerable uncertainty about the use of pneumococcal vaccine (ACIP 1981; Lancet 1981; Hirschmann and Lipsky 1981). The vaccine may benefit children with sickle-cell disease and nephrosis and persons with surgical or functional asplenia. It may also be helpful in patients with Hodgkin's disease, although the timing of immunization in this and other immunologic disorders may be critical in determining its effectiveness. In the United States recommendations suggest that pneumococcal immunization should also be considered for persons with chronic cardiopulmonary disease, hepatic and renal dysfunction, or diabetes mellitus, and perhaps for elderly persons (ACIP 1981). Although cost-effectiveness and cost-benefit studies in the United States have suggested the value of immunizing persons aged 65 years or older (Willems, Sanders, and Riddiough 1980; Patrick and Woolley 1981), in the United Kingdom, such a policy is not yet regarded as justified (Lancet 1981).

The current definition of high-risk groups for pneumococcal immunization is based on age and underlying medical conditon alone. There are no estimates of the numbers of persons at high risk in the United Kingdom. In the United States, this population may number 40 to 60 million persons (US Dept of Health, Education, and Welfare 1979). Thus, there is an obvious need for more precise criteria that can be used to determine who, among this group, are truly at high risk of developing serious or fatal pneumococcal pneumonia.

Almost 20 years ago, studies in the United States revealed that 20 to 30 per cent of persons who died of pneumonia and influenza had been discharged from hospital in the year before death (US Dept of Health, Education, and Welfare 1965a and b). These findings suggest that previous hospital discharge may be a useful marker for identifying many persons who are at increased risk of being hospitalized for or dying of pneumococcal pneumonia. Since vaccine-induced antibody levels—and presumably protection—have been shown to persist five to eight years after pneumococcal immunization (Hilleman, Carlson, McLean *et al*. 1981; Heidelberger, Dilapi, Siegel *et al*. 1950), these studies also suggest that, if a substantial portion of those who develop serious or fatal pneumonia can be shown to have been discharged from hospital within this period, a strategy to immunize patients with pneumococcal vaccine at the time of hospital discharge might be beneficial. With this in mind, the files of the Oxford Record Linkage Study have been examined to determine the patterns of previous hospital care among persons hospitalized for or dying of pneumonia to consider the implications of the results for a hospital-based approach to pneumococcal immunization.

Methods

The files of the Oxford Record Linkage Study contain abstracts of all hospital discharge summaries together with records of all births, stillbirths, and deaths occurring in a defined population. Each abstract contains identifying, demographic, social, and clinical data. The records are arranged so that all hospitalization and vital events relating to each person are linked together in temporal order to form a cumulative personal medical record. The data used for this study were collected for the period 1963–70 for a mixed urban and rural population of 336 000 to 386 000 people living in the county of Oxfordshire, England. The methods of collecting and linking these data have been described elsewhere (Acheson 1967; Baldwin, 1973).

Historical study

All persons hospitalized for or dying of pneumonia (*International Classification of Diseases*, ed 8 *ICD-8* 480–486) were determined for the year 1970. Both primary and secondary diagnoses were included. Patients were grouped by age, type of pneumonia, and outcome. For those who died, the place of death was noted, as was the duration of stay for those who died in hospital. The sequential record of each patient was examined for all previous episodes of hospital care up to two weeks before hospital admission for pneumonia or death outside hospital because of pneumonia. Conditions listed on previous discharge summaries were classified as high risk or non-high risk. High-risk conditions were defined as heart disease (*ICD-7* 410–447; *ICD-8* 393–429), lower respiratory tract disease (*ICD-7* 241, 480–502, 518–526, 527.1; *ICD-8* 466, 470–474, 480–486, 490–493, 510–518), cerebrovascular disease (*ICD-7* 330–334; *ICD-8* 430–438), and diabetes mellitus (*ICD-7* 260; *ICD-8* 250). Whenever a continuous period of care occurred in two or more hospitals, it was regarded as only one episode of care.

Cohort study

The files of time-sequenced linked data containing abstracts of hospital discharge records and death records were arranged in a series of sex-specific and five-year age-specific cohorts of persons discharged alive with any condition or with high-risk conditions only. Cohorts were selected on the basis of primary diagnosis only, in common with most other studies, but this excluded some patients with high-risk conditions that were not recorded as primary. For each cohort, the probability was determined of a person within the cohort being readmitted with or dying of pneumonia within specified time intervals up to eight years after discharge. A readmission was defined as any separately recorded episode of inpatient care and, therefore, included transfers to another hospital in the course of care or to convalescent units. Deaths included those occurring in hospital and those occurring at home or elsewhere. Separate analyses were conducted for persons who died whether or not they had been readmitted, and for the combined groups of persons who either were readmitted or died without readmission.

Cohort analysis

In the computer analysis, cohorts were defined as groups of persons satisfying sex, age, and diagnostic criteria (C) and outcome events (E) occurring within time k (in days, weeks, months, or years). Then: let i be the date on which criteria C were satisfied; a, the age of the patient at i; s, the sex of the patient; j, the date on which the patient experienced either outcome E or death from any other cause; and n, the duration of the data file (i.e. the length of time from the start of the data file to the date of the last entry on the data file).

For each patient satisfying criteria C: if event E occurs subsequent to C, compute $k=j-i$ and add 1 to the matrix B_{kas}; if event E does not occur, but death from another cause occurs, compute $k=j-i$ and add 1 to the matrix D_{kas}; if neither event E nor death from other causes occurs, compute the length of time k from i to the end of the data file and add 1 to the matrix P_{kas}.

Compute the matrix R_{kas} such that:

$$r_{kas} = 1 - \sum_{h=1}^{k} m_{has}$$

where M_{has} is a known matrix giving the proportion migrating out of the population at risk in month h. Then compute the matrix T_{uas} where:

$$t_{uas} = \sum_{h=u}^{n} [r_{pas}\, p_{has} + d_{has} + b_{has}] \tag{6.1}$$

which gives an estimate of the number of persons in each sex-age subgroup of the cohort followed up to u months after i, the date on which criteria C were satisfied.

The statistic b_{vas}/t_{vas} (6.2) gives the proportion of vulnerable patients with C who have E in month v, and:

$$\left[1 - \prod_{v=1}^{u} \left(1 - \frac{b_{vas}}{t_{vas}} \right) \right] \tag{6.3}$$

gives the proportion of all patients with C who have E sometime before the end of u months.

The cohort analysis accounted for varying periods at risk and corrected for cohort losses from migration during the period of follow-up. The proportions of persons migrating over time were estimated from a limited study of re-registrations of patients of general practitioners in the Oxfordshire Executive Council area in 1971 and were, therefore, approximations.

Expression 6.2 was summed to $v=u$ instead of expression 6.3 being calculated, and in the last term in formula 6.1, the summation started at the beginning of the time period. These approximations had the net effect of slightly reducing the probabilities of readmission and death, the maximum underestimate being 4 per cent at five years. The network of reporting hospitals in the Oxford Record Linkage Study includes hospitals outside Oxfordshire that are used by Oxford residents. These persons were included in the analysis. Conversely, persons living outside Oxfordshire but hospitalized in Oxfordshire were excluded from the analysis.

The probabilities obtained from the cohort analysis were used to estimate the number of persons who would have had to be immunized at the time of hospital discharge to prevent one subsequent readmission, death, or readmission and death combined, caused by preventable pneumococcal pneumonia. This number was calculated as follows: (100/per cent risk) × (100/per cent preventable with pneumococcal vaccine).

Results

Historical study

In the 1970 analysis, there were 793 persons for whom pneumonia was reported as the primary or underlying cause of hospital admission or death (Table 6.1). Pneumococcal infection was regarded as the cause for 28 per cent of the 364 persons who survived hospitalization for pneumonia but for only 8 per cent of the 429 deaths associated with pneumonia in or out of hospital. Of the 34 deaths attributed to pneumococcal pneumonia, only one occurred among persons younger than 45 years. For most pneumonias, the causative agent was not reported. This was particularly evident among persons aged 65 years or older, where 74 per cent of those who survived and 92 per cent of those who died had pneumonia caused by unspecified organisms (*ICD-8* 485, 486). Very few pneumonias were reported in other categories (*ICD-8* 480, 482–484).

Overall, 36 per cent of those who survived and 49 per cent of those who died had been discharged from hospital on one or more occasions within the previous five years. The proportions were only slightly higher for discharges within eight years of an admission for or death caused by pneumonia. The eight-year findings will not be discussed further. For patients older than 64 years, previous hospital care had little bearing on the outcome; 83 (48 per cent) of 172 patients who survived and 176 (51 per cent) of 357 patients who died had at least one discharge from hospital within the preceding five years ($\chi^2 = 0.017$, $P > 0.1$). Many persons with previous episodes of hospital care had been discharged with high-risk conditions. Among those older than 44 years who died, 20 to 24 per cent had been discharged with a high-risk condition. In addition, in the two groups of patients older than 64 years who

Table 6.1 Aetiology and patterns of previous hospital care among persons with hospital admission for or death caused by pneumonia. Oxford Record Linkage Study, 1970.

Age group (yrs.)	Outcome	Total No. of patients with pneumonia	Aetiology of Pneumonia			Percentage with hospital care within previous 5 yrs	
			Pneumococcal (ICD-8* 481) No.(%)	Organism unspecified (ICD-8 485, 486) No. (%)	Other (ICD-8 480, 482–484) No. (%)	All conditions	High-risk conditions†
0–4	Survived‡	57	10(18)	44(77)	3(5)	19	7
	Died§	9	0(0)	7(78)	2(22)	0	0
5–14	Survived	27	12(44)	13(48)	2(7)	22	4
	Died	1	1(100)	0…	0…	0	0
15–44	Survived	29	6(21)	19(66)	4(14)	28	7
	Died	7	0…	7(100)	0…	29	14
45–64	Survived	79	33(42)	44(55)	2(3)	29	9
	Died	55	7(13)	47(85)	1(2)	58	24
65–74	Survived	78	23(29)	54(69)	1(1)	51	18
	Died	84	7(8)	76(90)	1(1)	51	20
≥75	Survived	94	19(20)	74(79)	1(1)	46	26
	Died	273	19(7)	253(93)	1(0.4)	49	24
All ages	Survived	364	103(28)	248(68)	13(4)	36	14
	Died	429	34(8)	390)91)	5(1)	49	22
Total	…	793	137(17)	638(80)	18(2)	43	19

* ICD-8 indicates *International Classification of Diseases* (edition 8).
† See text for definition of high-risk conditions.
‡ Hospitalized for pneumonia and discharged alive.
§ Died of pneumonia in or out of hospital.

survived hospitalization, the proportions were 18 and 26 per cent, respectively.

The proportions of patients with pneumococcal pneumonia and with all other pneumonias who had had previous hospital care were compared for persons aged 45 years or older. For persons discharged with any condition and for those discharged with high-risk conditions, and for both outcomes, there were no significant differences at the 5 per cent level (Fisher's exact test), with one exception. Among persons aged 65 to 74 years who survived, seven (30 per cent) of 23 with pneumococcal pneumonia and 33 (60 per cent) of 55 with all other pneumonias had been discharged with any condition within the previous five years ($\chi^2 = 4.55$, $P < 0.05$).

Information on duration of stay was available for all but 11 of the 322 persons aged 45 years or older who died in a hospital. Death occurred within five days of admission in 37 (30 per cent) of 123 persons aged 45 to 74 years and in 51 (26 per cent) of 200 of those older than 74 years. An episode of previous hospital care was noted in 54 and 47 per cent of the patients in these two groups.

Cohort study

In the cohort study, three subsequent outcomes associated with pneumonia were considered: death alone, readmission to hospital alone, and re-admission and death combined. The cumulative probability of a person within a cohort experiencing each outcome was determined at six-month intervals up to eight years after hospital discharge with all conditions and with high-risk conditions. Separate calculations were obtained for five-year age cohorts up to 96 years of age and older. The calculated cumulative probabilities for cohorts younger than 40 years were very small, and those for cohorts 86 years or older were very large. They will not be considered further. The size of the intervening cohorts is given in Table 6.2.

An example of the analysis for one of the cohorts, the group of persons aged 66–70 years, is shown in Fig. 6.1. The probability of dying of pneumonia within five years of hospital discharge with any condition was 2.6 per cent and rose to 4.3 per cent for discharge with a high-risk condition. During the same interval, the probability of being readmitted to hospital for pneumonia was 4.8 per cent for persons previously discharged with any condition and 10.5 per cent for those discharged with high-risk conditions. At eight years, the probabilities were only slightly greater than those at five years.

Table 6.3 gives the probabilties of death, readmission, and readmission and death combined because of pneumonia five years after hospital discharge for five-year age cohorts. At all ages, the probabilities for high-risk persons exceeded those for persons discharged with all conditions. The relative risks are also given for those discharged with high-risk conditions

Table 6.2 Size of cohorts in cohort analysis.
Oxford Record Linkage Study, 1963–70.

Entering cohort (yrs)	All conditions No.	High-risk conditions No. (%)
41–45	4 953	282 (5.7)
46–50	4 616	394 (8.5)
51–55	5 196	692 (13.3)
56–60	4 999	844 (16.9)
61–65	4 654	937 (20.1)
66–70	3 639	862 (23.7)
71–75	3 284	870 (26.5)
76–80	2 774	866 (31.2)
81–85	1 862	576 (30.9)
41–85	35 977	6 323 (17.6)
All ages	83 815	8 808 (10.5)

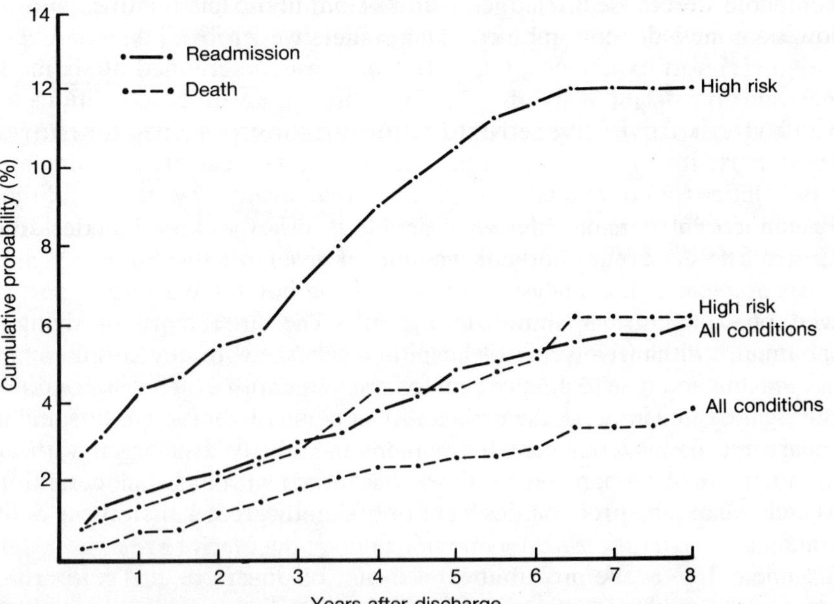

Fig. 6.1 Cumulative probability of readmission and of death caused by pneumonia after previous hospital discharge for persons aged 66 through 70 years: Oxford Record Linkage Study 1963–70. See text for definition of high risk and all conditions.

Table 6.3 Probability and relative risk of death, readmission, and readmission and death caused by pneumonia five years after hospital discharge. Oxford Record Linkage Study, 1963–70.

Entering cohort (yrs)	Death		Readmission		Readmission and death	
	All conditions P^*	High-risk conditions P (RR)†	All conditions P	High-risk conditions P (RR)	All conditions P	High-risk conditions P (RR)
41–45	0.1	ND‡	0.8	3.4 (4.3)	0.8	3.4 (4.3)
46–50	0.2	0.6 (3.0)	0.9	5.4 (6.0)	0.9	5.8 (6.4)
51–55	0.3	0.5 (1.7)	1.5	4.0 (2.7)	1.7	4.3 (2.5)
56–60	0.5	1.2 (2.4)	1.9	4.8 (2.5)	2.3	5.8 (2.5)
61–65	1.5	3.6 (2.4)	3.2	7.9 (2.5)	4.2	10.2 (2.4)
66–70	2.6	4.3 (1.7)	4.8	10.5 (2.2)	6.2	12.5 (2.0)
71–75	3.7	6.3 (1.7)	6.1	11.3 (1.9)	8.4	15.3 (1.8)
76–80	7.1	8.3 (1.2)	10.0	16.3 (1.6)	14.0	19.8 (1.4)
81–85	10.0	12.3 (1.2)	11.0	13.2 (1.2)	17.3	21.0 (1.2)

$*P$ indicates probability expressed as percentage.
†RR indicates relative risk i.e.,P (high -risk conditions)/P (all conditions).
‡ ND indicates not determined for cohorts with fewer than four deaths.

compared with those discharged with all conditions. The relative risks were lower among older cohorts, but in all instances were greater than one.

Estimated preventive effect of immunizing persons discharged from hospital

Pneumococcal vaccine was not available during the period under study. Given the risk of an episode of pneumonia developing within five years of hospital discharge, it was possible to estimate the number of persons who would have had to be immunized (assuming vaccine had been available) at the time of discharge to prevent one pneumococcal event. Such estimates assume that the distribution of preventable pneumococcal pneumonias was the same as that of all pneumonias. Since the percentage of all pneumonias caused by pneumococcal serotypes included in the vaccine was unknown, a broad range of estimates was calculated. In this sensitivity analysis, it was assumed that 15, 30, or 45 per cent of pneumonia readmissions or deaths could have been prevented by immunization at the time of previous hospital discharge. If it is assumed that 80 per cent of pneumococcal pneumonias were caused by vaccine serotype organisms and that the vaccine was 80 per cent effective in preventing these events, these figures correspond to estimates that 23, 47, and 70 per cent of all pneumonias were caused by pneumococcal organisms of any serotype.

Figures 6.2 and 6.3 show the numbers of persons discharged with all conditions and with high-risk conditions who would have had to be immunized on discharge to prevent each outcome event within the next five years. They suggest that, regardless of whether 15, 30, or 45 per cent of pneumonias is assumed to have been preventable with vaccine, relatively few discharged patients would have had to be immunized to prevent each subsequent hospital readmission, death, or readmission and death combined because of pneumococcal pneumonia. If deaths in the cohort of patients aged 66–70 years are examined, immunization of between 257 to 85 persons previously discharged with any condition and between 155 to 52 persons discharged with high-risk conditions would have prevented each death caused by pneumococcal pneumonia if 15–45 per cent of all pneumonia-related deaths had been preventable with pneumococcal vaccine. If re-admissions and deaths in this cohort are considered together, immunization

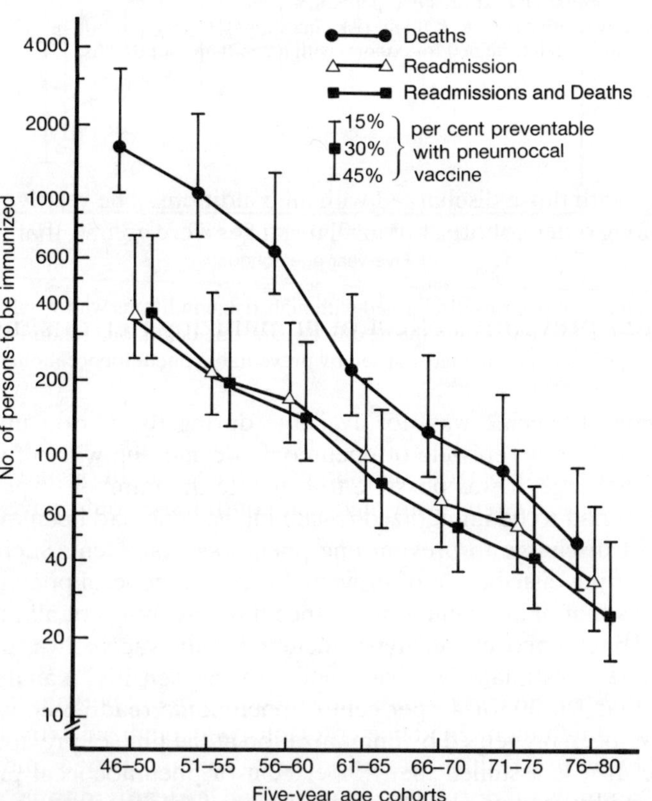

Fig. 6.2 Number of persons discharged with all conditions who would have had to be immunized to prevent, with the next five years, one death, one readmission, and one readmission and death combined, caused by preventable pneumococcal pneumonia.

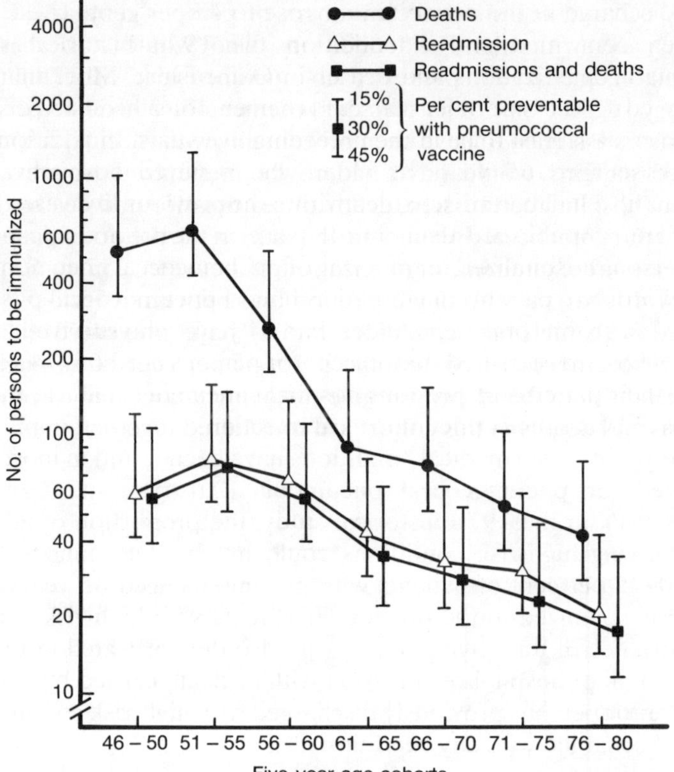

Fig. 6.3　Number of persons discharged with high-risk conditions who would have had to be immunized to prevent, within the next five years, one death, one readmission, and one readmission and death combined, caused by preventable pneumococcal pneumonia.

of between 108 to 36 persons discharged with any condition and between 53 to 18 persons discharged with high-risk conditions would have prevented each event.

Comment

The US study in the early 1960s indicated that many persons who died of pneumonia and influenza had been previously hospitalized in the last year of life (US Dept of Health, Education, and Welfare 1965a and b). More recently, a study of pneumonia and influenza deaths during periods of epidemic influenza has shown that 39 per cent occurred among persons discharged within the previous year (Barker and Mullooly 1982). During a comparable period when influenza was not present, the proportion with hospital discharge in the previous year rose to 47 per cent. These findings have been confirmed and extended in the 1970 historical study of pneumonia hospitalizations and deaths in Oxfordshire. More than half of persons aged 45 years or older who died of pneumonia had been discharged from a hospital at least once in the preceding five years. In addition, almost half of those aged 65 years or older who survived hospitalization for pneumonia also had been discharged within the previous five years.

There are no published data on the patterns of previous hospital care among persons hospitalized for or dying of pneumococcal pneumonia alone. In the Oxfordshire patients diagnosed as having pneumococcal pneumonia, analysis of all groups of patients older than 44 years who survived hospitalization or who died disclosed that, except for persons aged 65–74 years who survived, their patterns of previous hospital care were similar to those who diagnoses did not specify pneumococcal infection.

A variety of chronic medical conditions have been found in most patients hospitalized for pneumococcal pneumonia (Austrian and Gold 1964; Mufson 1981). In the 1970 historical study, the proportion of all patients with underlying high-risk conditions could not be determined. Approximately 20–25 per cent of patients with pneumonia aged 65 years or older had been hospitalized in the previous five years with high-risk conditions. The importance of underlying illness was also demonstrated in the cohort study. The risk of hospital readmission with or death caused by pneumonia was greater for persons previously discharged with high-risk conditions than for those discharged with any condition.

The benefits of pneumococcal immunization could be anticipated if the incidence of pneumococcal pneumonia were known. In the United States, crude estimates from two population-based studies have suggested rates of 2–6 per 1000 adults per year (Foy, Wentworth, Kenny *et al*. 1975; Austrian 1975). A recent population-based study has reported an overall incidence of pneumococcal bacteraemia of only 8.5 per 100 000 persons per year (Filice, Darby, and Fraser 1980). In out-patient settings, pneumococcal infections may account for only 15–20 per cent of all adult pneumonias (Foy, Wentworth, Kenny *et al*. 1975), but in the hospital setting they constitute a substantially larger proportion of all pneumonias. Studies from ten US hospitals indicate that 30–67 per cent of pneumonias may be caused by pneumococcal infection (Mufson 1981). These rates reflect experience in teaching hospitals that generally admit only patients with serious disease. The same rates might not be observed in smaller community hospitals.

In the 1970 study of pneumonias, 28 per cent of all who survived hospitalization for pneumonia were diagnosed as having pneumococcal pneumonia, and in persons aged 45–64 years and 65–74 years, the percentages were 42 and 29 per cent, respectively. For those who died of pneumonia, only 8 per cent were given the same diagnosis. Experience in some US hospitals has shown that the coded diagnosis of pneumococcal pneumonia may be unreliable (Saitz 1978). However, a number of unpublished studies of the reliability of the Oxford Record Linkage Study abstracts indicate that the coding was likely to have been accurate, although the rubric for pneumococcal pneumonia (*ICD-8* 481) includes fibrinous, lobar, and massive pneumonias for which the causative agent is not specified. Conversely, in almost all other patients, neither the type nor the causative agent was specified. In persons aged 45–74 years who survived, it is reasonable to assume that approximately 30–40 per cent had pneumococcal pneumonia, and similar proportions in other age groups and in those who died may have experienced pneumococcal infection. These estimates are derived from the Oxfordshire area as a whole and not just the large teaching institutions.

The cohort study provided estimates of the probability of readmission to hospital for or death caused by pneumonia among persons discharged within the previous five years. The resulting probabilities were then used to calculate the number of persons who would have had to be immunized with pneumococcal vaccine at the time of discharge to prevent a subsequent hospitalization and/or death caused by preventable pneumococcal pneumonia. If only 15 per cent of pneumonia readmissions and deaths combined had been preventable with vaccine, during the next five years, one of these events could have been prevented by immunizing fewer than 160 persons older than 60 years discharged with all conditions and fewer than 120 to 160 persons older than 45 years discharged with high-risk conditions. If 30 per cent of these events had been preventable, the number of persons who would have had to be immunized is reduced by half. On the other hand, if the percentage of pneumonias that was caused by pneumococcal infection had been less than what has been previously reported from hospitals (Mufson 1981), or if the efficacy of the vaccine had been less than 80 per cent, the number of persons who would have had to be immunized to prevent each outcome event would have been greater. Nevertheless, even it if is cautiously assumed that only 7.5 per cent of these events could have been prevented, in persons older than 60 years discharged with any condition, immunizing 320 or fewer persons would have prevented each pneumococcal pneumonia readmission or death. For persons discharged with high-risk conditions, each outcome event would have been prevented by immunizing 230 to 310 persons aged 45–60 years and 130 or fewer persons older than 60 years. The sensitivity analysis demonstrates that, regardless of which percentage of pneumonia readmissions and deaths is estimated to have been preventable by immunization, each event could have been prevented by immunizing a few hundred (and in older patients, often fewer than 100) discharged patients. With numbers in this range, the estimates are relatively insensitive to assumptions regarding which proportion of all pneumonia events may have been preventable. In older patients, the estimates are also relatively insensitive to whether patients were previously discharged with high-risk conditions.

In the historical study, the proportions of patients with pneumonia who received hospital care within the previous eight years were only slightly higher than those within the previous five years. In the cohort study, the five- and eight-year probabilities for subsequent readmission for or death caused by pneumonia were also similar. Whether immunity after pneumococcal immunization lasts eight years or only five is not yet known. If it is found that immunity lasts for only five years, the findings suggest that immunization of discharged persons has the potential for reaching almost all of those discharged persons who will be readmitted or die of pneumonia within the next eight years.

The chief approach to controlling pneumococcal infections relies on anti-microbial therapy. Unless or until pneumococcal organisms resistant to multiple antimicrobial agents become widely prevalent, it is unlikely that this approach will be superseded by one that focuses on primary prevention. Yet it is known that death rates from pneumococcal pneumonia occurring within five days of onset of illness may not be reduced by antimicrobial therapy (Austrian and Gold 1964). The 1970 historical data show that one-fourth to one-third of pneumonia deaths occurred within the first five days of hospitalization. Approximately half of these patients had been discharged within the previous five years.

In the United Kingdom, the traditional approach to immunization against another major adult respiratory tract infection, influenza, has emphasized the importance of the general practitioner (The Standing Medical Advisory Committee for The Central Health Services Council and The Minister of Health 1963). In the United States, 80 per cent of influenza immunizations have been given in physicians' offices (US Dept of Health, Education, and Welfare 1975). There are no data that describe the use of pneumococcal vaccine by general practitioners in the United Kingdom and only limited data regarding its use by generalist physicians in the United States (Pantell and Stewart 1979). In addition, nothing is known of the proportion of patients who have attended a general practitioner's surgery in the United Kingdom or a physician's office in the United States within five years of developing pneumococcal infection. In all likelihood, a substantial majority has done so. It would be important to know whether general practitioners and generalist physicians are immunizing elderly and high-risk patients, because immuniza-tion in the primary care setting would be more likely to reach a larger number of persons destined to experience pneumococcal infections. Limited data from the United States suggest that pneumococcal vaccine has not been widely used by office-based physicians (Hilleman, Carlson, McLean *et al.* 1981; Pantell and Stewart 1979).

An alternative and complementary approach to office-based pneu-mococcal immunization might be to offer vaccine at the time of hospital discharge. It would be helpful to know the relative advantage of immunizing discharged patients compared with immunizing all persons with high-risk conditions (whether or not they had been discharged), or all persons in general. It might be possible to estimate the efficacy of these strategies by mounting cohort studies, but they would be costly and would be justified only if immunization can be shown to be effective on a more limited scale. It should be emphasized that, for each age group in any given year, the number of persons discharged from hospital is substantially smaller than the total number of persons in the group as a whole or the number of persons with high-risk conditions. The results of this study suggest that if current patterns of previous hospital care among persons with pneumonia are similar to those found in Oxfordshire in 1970, it may be expected that almost 50 per cent of pneumonia-related deaths preventable with pneumococcal vaccine will occur among the smaller percentage of persons with a recent episode of hospital care, and 20–25 per cent of such deaths will occur among the very small proportion of the population previously discharged with high-risk conditions.

Acknowledgements

The senior author wishes to dedicate this article to the memory of Dr Baldwin.

The clerical staff of the Oxford Record Linkage Study prepared data; the computing staff of the Unit of Clinical Epidemiology, University of Oxford, conducted computer analyses, Sarah Euren assisted with statistical work, and Evelyn Clarke and Susan Olsen prepared the manuscript.

References

ACHESON, E. D. (1967). *Medical Record Linkage*. Oxford University Press, Oxford, England.

ACIP (1981). Recommendations of the Immunization Practices Advisory Committee. Pneumococcal polysaccharide vaccine. *MMWR* **30**, 410–19.

AUSTRIAN, R. (1975). Random gleanings from a life with the pneumococcus. *J. Infect. Dis.* **131**, 474–84.

—— (1980). The assessment of pneumococcal vaccine. *N. Engl. J. Med.* **303**, 578–80.

—— AND GOLD, J. (1964). Pneumococcal bacteremia with especial reference to bacteremic pneumococcal pneumonia. *Ann. Intern. Med.* **60**, 759–76.

——, DOUGLAS, R. M., SCHIFFMAN, G. *et al.* (1976). Prevention of pneumococcal pneumonia by vaccination. *Trans. Assoc. Am. Physicians* **89**, 184–94.

BALDWIN, J. A. (1973). Linked record medical information systems. *Proc. R. Soc. Lond.* **184**, 403–20.

BARKER, W. H. AND MULLOOLY, J. P. (1982). Pneumonia and influenza deaths during epidemics: implications for prevention. *Arch. Intern. Med.* **142**, 85–9.

BROOME, C. V., FACKLAM, R. R., AND FRASER, D. W. (1980). Pneumococcal disease after pneumococcal vaccination: an alternative method to estimate the efficacy of pneumococcal vaccine. *N. Engl. J. Med.* **303**, 549–52.

FILICE, G. A., DARBY, C. P., AND FRASER, D. W. (1980). Pneumococcal bacteremia in Charlestown County, South Carolina. *Am. J. Epidemiol.* **112**, 828–35.

FOY, H. M., WENTWORTH, B., KENNY, G. E. *et al.* (1975). Pneumococcal isolations from patients with pneumonia and control subjects in a prepaid medical care group. *Am. Rev. Respir. Dis.* **111**, 595–603.

HEIDELBERGER, M., DILAPI, M. M., SIEGEL, M. *et al.* (1950). Persistence of antibodies in human subjects injected with pneumococcal polysaccharides. *J. Immunol.* **65**, 535–41.

HILLEMAN, M. R., CARLSON, A. J. JR., McLEAN, A. A. *et al.* (1981). *Streptococcus pneumoniae* polysaccharide vaccine. Age and dose responses, safety, persistence of antibody, revaccination, and simultaneous administration of pneumococcal and influenza vaccine. *Rev. Infect. Dis.* **3**, S31–S42.

HIRSCHMANN, J. V. AND LIPSKY, B. A. (1981). Pneumococcal vaccine in the United States. A critical analysis. *JAMA* **246**, 1428–32.

Lancet (1981). Indications for pneumococcal vaccine (editorial). **1**, 251–3.

MUFSON, M. A. (1981). Pneumococcal infections. *JAMA* **246**, 1942–8.

Office of Population Censuses and Surveys (1980). *Mortality Statistics: Cause—1978.* Series DH2 No. 5.

PANTELL, R. H. AND STEWART, T. J. (1979). The pneumococcal vaccine. Immunization at a crossroad. *JAMA* **241**, 2272–4.

PATRICK, K. M. AND WOOLLEY, F. R. (1981). A cost-benefit analysis of immunization for pneumococcal pneumonia. *JAMA* **245**, 473–7.

SAITZ, E. W. (1978). Pneumococcal pneumonia: a misleading diagnosis for audit studies. *JAMA* **239**, 2372.

The Standing Committee for The Central Health Services Council and the Minister of Health (1963). *Active Immunization Against Infectious Disease.* March, 1963.

TURK, D. C. (1978). Frequencies of pneumococcal types causing serious infections in patients admitted to the Radcliffe Infirmary, Oxford, 1969–77. *J. Hyg. Camb.* **81**, 227–38.

United States Department of Health, Education, and Welfare (1965a). *Hospitalization in the Last Year of Life—United States 1961.* Series 22, No. 1, table 2. National Center for Health Statistics.

—— (1965b). *Episodes and Duration of Hospitalization in the Last Year of Life: United States—1961.* Series 22, No. 2, tables 6 and 10. National Center for Health Statistics.

—— (1975). *United States Immunization Survey, 1974.* Publication (CDC) 75-8221. Centers for Disease Control, Atlanta.

—— (1979). *United States Immunization Survey, 1978.* Publication (CDC) 79-8221. Centers for Disease Control, Atlanta.

WILLEMS, J. S., SAUNDERS, C. R., RIDDIOUGH, M. A. *et al.* (1980). Cost effectiveness of vaccination against pneumococcal pneumonia. *N. Engl. J. Med.* **303**, 553–9.

7

Schizophrenia and physical disease: a preliminary analysis of the data from the Oxford Record Linkage Study

THE LATE J. A. BALDWIN

The study reported in this chapter originated from a request by the World Health Organization to participate in the Study of Determinants of Outcome of Severe Mental Disorders, which it is co-ordinating as a multinational study in succession to the International Pilot Study of Schizophrenia. One sub-study is of the General Morbidity of Psychiatric Patients and their Families and is based on psychiatric case registers in Denmark, Hawaii, Monro County, N.Y., Moscow, Nagasaki, and Oxford. The Oxford study described here was a preliminary analysis of available data as a guide to design of specific studies in the other centres. The results are therefore tentative and subject to revision upon further examination of the data.

Background

A review of literature on the relation between physical disease and schizophrenia showed that a large number of disorders have been postulated to be associated either positively or negatively (antagonistically) with schizophrenia at various times throughout the last two millenia, yet there are very few firmly established conclusions (Baldwin 1979). Most sound observational studies have been of mortality while the adequacy of design and scale of many studies of both mortality and morbidity are questionable. There is therefore a need for a systematic statistical evaluation of the range of illness found in association with schizophrenia in order to concentrate attention on relationships likely to be of aetiological relevance. The requirement is for valid estimates of the *relative risk* of diseases in schizophrenics in comparison with other psychiatric disorders and with the general population from which the patients come. The collection of cumulative personal medical records in the Oxford Record Linkage Study is a convenient source of information for this purpose.

Reprinted from the *The Biochemistry of Schizophrenia and Addiction* (Ed. Gwynneth Hemmings) Lancaster, England, MTP, (1980).

Material and Method

The Oxford Record Linkage Study is a collection of abstracts of clinical records of all types of hospitalization together with records of all births, still-births, and deaths occurring in a defined population at risk. The abstract records contain identifying, demographic, social, and clinical data and are arranged so that all the hospitalization and vital events relating to an individual are linked together in temporal order to form cumulative personal medical records. This medical information system was started in 1963 and continues to the present day. The techniques of collecting and linking the data have been described elsewhere (Acheson 1967; Baldwin 1973). Data used for the study reported in this chapter were those for the nominal 8-year period from 1963 to 1970 in respect of the County of Oxfordshire and part of the County of Berkshire, representing a mixed urban and rural community of between 750 000 and 800 000. The whole of this geographical area was not covered throughout the 8-year period and the mean period at risk was approximately 3.25 years. The total number of hospitalization, birth, and death records was 559 266, and when linked together formed 366 862 cumulative personal medical records. The diagnostic information consisted of up to two diagnoses for each hospitalization or vital event, which were coded according to the International Classification of Diseases (Seventh and Eighth Revisions). There were 2314 persons diagnosed schizophrenic. Classification to fourth digit level was used wherever numbers permitted (Table 7.1).

The objective of the study was to estimate whether schizophrenia was preceded or followed by another disease more often than would be expected by chance, making use of all the available diagnostic information from the cumulative personal medical records. Observed numbers of individuals having each disease studied prior to and following schizophrenia were

Table 7.1 Sub-types among 2314 persons diagnosed as having schizophrenia during the period 1963–70

Simple	154
Hebephrenia	82
Catatonia	42
Paranoid	375
Acute	83
Latent/residual	18
Schizo-affective	72
Other/not otherwise stated	1 720

Note: The total is greater than 2314 because some individuals were diagnosed as in different sub-types at different times.

obtained, omitting only repeated diagnostic information. Expected numbers were derived from the sex and age-specific incidence rates, for the calendar period at risk, of probands, using the linked data as numerators and community population at risk as denominators, adjusted for period at risk, variation of risk with time (secular, seasonal, and epidemic trends), sex differences in risk, the cohort effect on age, population migration out of the area, and mortality. The statistical method is described in Chapter 3.

Results were expressed as relative risks (ratio of observed to expected numbers) tested for statistically significant deviations from unity. Expectations were considered to be distributed as Poisson variables and exact probabilities of differences from observed values calculated. Except where differences in relative risks for each temporal order of associated diseases appeared likely to be both important and reliable, results for both temporal orders were combined. Simultaneously occurring diagnoses are difficult to handle statistically and were omitted from the main analyses. Where numbers of such cases altered the result significantly the effect is reported and discussed in the text.

Similar calculations were obtained in respect of individuals with the diagnosis of affective psychosis (1489 persons), to ascertain the degree of association of each disease with another functional psychosis, and non-psychotic depression (3915 persons), to ascertain the degree of association with a non-psychotic psychiatric disorder. Using the same method of calculating relative risks for all three psychiatric conditions ensured strict comparability between the findings for schizophrenia and the controls, permitting an estimate of the degree of specificity of any given association with schizophrenia.

The study was intended only as a preliminary analysis to guide further work and the limitations of the data and of the method must be stated at the outset. Relatively small numbers of cases of schizophrenia and controls as well as of many uncommon diseases were available so that the material was not adequate to assess a number of potentially important associations. Where small numbers are involved, even infrequent errors in the clinical data or in the coding and processing of it may cause unreliable results. Results based on small numbers should be verified by reference to original records but this was not practicable with the time and resources allocated for the study. The nominal period of 8 years with a mean period at risk of only 3.25 years was too short to enable conclusions to be drawn about relationships between diseases which are likely to occur at widely different ages in individuals.

Deficient ascertainment of morbidity is likely to have distorted some results in various ways. The fact that the collection of data did not commence until 1963 entails that diseases occurring in individuals before this year were incompletely ascertained. Only morbidity leading to admission to a hospital in-patient service or death (in hospital or community) was included so that physical disease treated wholly in mental hospitals or by means of out-patient or primary care was not recorded. A total of 328 schizophrenics were resident in hospital at the end of the follow-up period and had not been transferred to another facility for treatment of a physical disease.

The diagnostic information was in all cases obtained from clinical records. Thus there was likely to have been variation in diagnostic standards over the 8-year period, particularly in respect of psychiatric diagnoses. There are strong associations between the diagnoses of schizophrenia, affective psychosis, and depression, many individuals exhibiting features of more than one of these conditions in the course of successive hospital admissions.

The estimation of periods at risk is in some respects imprecise. For example, information on migration out of the population at risk was sparse, the weights given to this factor having been obtained from a single study covering a year following the end of the follow-up period.

These limitations are likely to have had conflicting and varied effects on the results obtained. Deficient ascertainment will have caused underestimation of associations in most instances and for this reason low relative risks should be interpreted with caution. In contrast, high relative risks are likely to be more reliable. Since both the schizophrenia group and the control diagnoses were, in the main, subject to the same constraints, greater confidence may be had in comparative results between these groups than in the absolute values of each group in isolation.

Results

Results are given for conditions claimed or hypothesized to be positively or negatively associated with schizophrenia in previous literature (Baldwin 1979), or found *de novo* in the course of the analysis.

Cancer

Summary results for some common cancers and for certain cancers noted in previous literature to be positively or negatively associated with schizophrenia are set out in Table 7.2. There were no robustly significant increases or decreases of relative risk for any cancer or all cancers in relation to either schizophrenia as a whole or the affective psychoses and non-psychotic depressions. There was a marginally significant excess of cancer of the stomach in non-specific schizophrenics and of cancer of the breast in hebephrenic women. A just significant increase in relative risk of cancer of the oesophagus was found in schizophrenics as a whole when simultaneously diagnosed cases were taken into account (relative risk $= 4.39$, $p = 0.03$). Although the period at risk was too short for confident evaluation, these results are in general accord with most comparable estimates. Nevertheless, the study did not take account of duration of hospitalization so that low relative risks for long-stay mental patients reported previously by Katz *et al.* (1967) may have been masked.

Table 7.2 Relative risk of specified cancers in schizophrenia cases and controls

Cancer	No. cases	Schizophrenia		Affective Psychosis		Depression	
		Observed No.	Relative Risk	Observed No.	Relative Risk	Observed No.	Relative Risk
Breast	2 438	7	0.96†	6	0.70	17	1.02
Colon/rectum	2 443	5	0.77	7	1.08	11	0.90
Haematopoetic	1 177	1	0.33	1	0.42	2	0.39
Lung/bronchus	3 052	5	0.64	6	1.07	17	1.38
Oesophagus	302	2	2.93	1	1.53	0	—
Pancreas	533	2	1.65	1	0.88	5	2.28
Stomach	1 312	5	1.73‡	4	1.46	6	1.16

† Marginally significant excess in hebephrenics.
‡ Marginally significant excess in non-specific schizophrenia.

Heart and vascular disease

Table 7.3 gives summary results for the commonest groups of circulatory and cardiovascular diseases. There was a significant excess of arteriosclerotic heart disease in both schizophrenics and control groups but in distinction from the controls there was no significant elevation of relative risk for any other group of circulatory diseases among schizophrenics.

Respiratory disease

There was a significant excess of pneumonia in relation to both schizophrenia and the controls, but in contrast to the control groups there was no excess of bronchitis among schizophrenics (Table 7.4). Pulmonary tuberculosis occurred more often than expected by chance prior to the diagnoses of schizophrenia and this was evident after exclusion of diagnoses of pleural effusion of presumed but unconfirmed tuberculous origin (Table 7.5). Non-psychotic depression was also more common than expected after pulmonary tuberculosis.

Epilepsy

Results for the main types of epilepsy are set out in Table 7.6. Significantly raised relative risks for major epilepsy, epileptic psychosis, and unspecified epilepsy were found among both schizophrenics and the control groups. For schizophrenics, all these excesses were limited to paranoid and non-specific schizophrenia. When simultaneously occurring diagnoses were included, the relative risk of non-specific epilepsy in non-specific schizophrenia was increased to 5.65 and of major epilepsy in paranoid schizophrenia to 24.69 though the validity of these ratios is somewhat uncertain.

Parkinson's disease

There were increases in the relative risks of schizophrenia, affective psychoses, and non-psychotic depression in the course of Parkinson's disease, but the excess was significant only in the last of these (Table 7.7). Parkinsonism was marginally significantly more frequent than expected by chance following all three groups of psychiatric disorder.

Asthma and hay fever

The incidence of allergic conditions was low, although within chance limits in schizophrenia but was significantly raised in both control groups.

Table 7.3 Relative risk of diseases of heart and circulatory system in cases of
schizophrenia and controls

Disease	No. cases	Schizophrenia		Affective psychosis		Depression	
		Observed No.	Relative Risk	Observed No.	Relative Risk	Observed No.	Relative Risk
Arterio-sclerotic heart disease	18 404	70	1.71***	60	1.55***	143	1.98***
Hypertensive heart disease	1 418	1	0.29	3	0.81	16	2.38**
Other hypertensive disease	2 895	9	0.96	12	1.36	41	2.31***
Arteriosclerosis	1 646	5	1.46	7	2.00	17	2.79***
Vascular lesions of CNS	11 550	31	1.26	45	1.64**	116	2.44***

** indicates probability $p < 0.01$.
*** indicates probability $p < 0.001$.

Table 7.4 Relative risk of pneumonia and bronchitis in schizophrenia and controls

Disease	No. cases	Schizophrenia		Affective Psychosis		Depression	
		Observed No.	Relative Risk	Observed No.	Relative Risk	Observed No.	Relative Risk
Pneumonia	10 770	34	1.63**	37	1.80***	100	2.64***
Bronchitis	9 068	23	1.05	33	1.78**	88	2.41***

** indicates probability $p < 0.01$
*** indicates probability $p < 0.001$

Table 7.5 Relative risk of pulmonary tuberculosis in cases of schizophrenia and controls

Disease	No. cases	Schizophrenia		Affective psychosis		Depression	
		Observed No.	Relative Risk	Observed No.	Relative Risk	Observed No.	Relative Risk
Before mental illness							
Pulmonary TB	1 885	8	2.93**	1	0.47	11	2.28*
Excl. pleural effusion	1 484	7	3.14**	1	0.59	10	2.52**
After mental illness							
Pulmonary TB	1 885	4	1.15	4	2.18	13	2.48**
Excl. pleural effusion	1 484	4	1.47	3	2.35	8	2.01

 * indicates probability $p < 0.05$
** indicates probability $p < 0.01$

Table 7.6 Relative risk of specified epilepsies in cases of schizophrenia and controls

Disease	No. cases	Schizophrenia		Affective psychosis		Depression	
		Observed No.	Relative Risk	Observed No.	Relative Risk	Observed No.	Relative Risk
Major epilepsy	201	3	5.32*†	3	10.75**	3	3.82*
Minor epilepsy	57	0	—	1	12.66	1	4.90
Focal epilepsy	73	0	—	1	8.62	1	3.28
Epileptic psychosis	46	5	32.68***†	0	—	4	17.78***
Epilepsy N.O.S.	1 888	14	2.76***†	10	3.17**	38	4.45***

† Significant excesses among paranoid schizophrenics and non-specific schizophrenics.
* indicates probability $p < 0.05$
** indicates probability $p < 0.01$
*** indicates probability $p < 0.001$

Table 7.7 Relative risk of Parkinson's disease in cases of schizophrenia and controls

Disease	No. cases	Schizophrenia		Affective psychosis		Depression	
		Observed No.	Relative Risk	Observed No.	Relative Risk	Observed No.	Relative Risk
Before mental illness	1 235	2	2.32	3	2.80	6	3.78**
After mental illness	1 235	5	2.57*	6	2.60*	9	2.19*

* indicates probability $p < 0.05$
** indicates probability $p < 0.01$

Peptic ulcer

The relative risk of peptic ulcer, including both gastric and duodenal types, was non-significantly low in schizophrenia, almost unity in affective psychosis, and marginally significantly raised in non-psychotic depression, this increase being limited to depression following the occurrence of ulcer (Table 7.8).

Diabetes mellitus

A marginally reduced relative risk was found among schizophrenics and there was a marginally raised risk of depression following diabetes mellitus (Table 7.8). Simultaneous diagnoses of schizophrenia and diabetes numbered seven, increasing the relative risk to 1.00. The chronicity of diabetes mellitus increases confidence in the validity of the combined figures and the initial reduced relative risk is almost certainly artefactual.

Malabsorption

Though the relative risk of malabsorption in schizophrenics was nearly threefold, this was within the limits of chance (Table 7.8). There was one additional case in which malabsorption and schizophrenia were diagnosed simultaneously, increasing the relative risk to 4.28 ($p = 0.034$). There was one case of hebephrenia following malabsorption, which was significantly more than expected. Depression in malabsorption was also three times as common as expected and this was a just significant excess. There were no cases of confirmed coeliac disease in this group of diagnoses.

Arthritis

There was a significant deficit of rheumatoid arthritis among schizophrenics as a whole (Table 7.8). A second unconfirmed case was found in whom the two conditions were diagnosed simultaneously increasing the relative risk to 0.30 ($p = 0.037$). The relative risk of osteoarthritis was also marginally significantly low in schizophrenia. The relative risk of both conditions was close to unity in both control groups.

Thyroid disorders

While the number of schizophrenics with thyrotoxicosis was similar to that expected, there was a fourfold increase in relative risk of myxoedema, which was just significant for schizophrenia as a whole. The excess was restricted to catatonics and non-specific schizophrenics and was significant in both groups (Table 7.8). One further case had both diagnoses simultaneously, raising the overall relative risk to 5.59 ($p = 0.002$). The risk of both

Table 7.8 Relative risk of certain diseases in cases of schizophrenia and controls

Disease	No. cases	Schizophrenia		Affective psychosis		Depression	
		Observed No.	Relative Risk	Observed No.	Relative Risk	Observed No.	Relative Risk
Asthma/hay fever	3 252	6	0.74	11	1.99*	34	2.46***
Peptic ulcer	3 973	11	0.76	10	1.01	31	1.40*
Diabetes mellitus	4 422	6	0.47*	14	1.09	37	1.52*
Malabsorption	266	2	2.85	1	1.81	4	3.06*
Rheumatoid arthritis	2 122	1	0.15**	10	1.38	13	0.94
Osteoarthritis	3 331	5	0.46*	12	1.02	23	1.09
Thyrotoxicosis	1 498	5	0.94	10	2.06*	25	2.18***
Myxoedema/cretinism	339	4	4.47*	4	3.95*	12	5.73***

   * indicates probability $p < 0.05$
  ** indicates probability $p < 0.01$
*** indicates probability $p < 0.001$

conditions was significantly raised among both affective psychoses and non-psychotic depressions, highly significant in respect of the latter.

Gallbladder disease

There was a highly significantly reduced relative risk of gallbladder disease among schizophrenics (Table 7.9). The risk was also low among affective psychotics but within chance limits, while among the non-psychotic depressions the observed and expected numbers were nearly equal.

Diseases of bone

Osteomyelitis and certain other diseases of bone were very significantly reduced in the group of schizophrenias (Table 7.9). There was a non-significantly low risk in the affective psychoses but not in the depressions.

Urinary diseases

A very significantly low relative risk of a group of urinary diseases, mainly calculi, was found among male schizophrenics only but not among males with affective psychoses or non-psychotic depressions (Table 7.10).

Head injury

There was a marginally significant increase in the relative risk of head injury in both male schizophrenics and male affective psychotics and a highly significant increase in association with depression (Table 7.10).

Appendicitis

Appendicitis was significantly uncommon in the female schizophrenics but not in the control groups (Table 7.11).

Varicose veins

There was a very significant deficit of varicose veins among female schizophrenics (Table 7.11), though the deficit was reduced by the addition of two simultaneously diagnosed cases (relative risk $= 0.36$, $p = 0.014$). The relative risk was also low though within chance limits in both control groups.

Menstrual and uterine disorders

There were low relative risks of menstrual disorders, benign uterine fibromas, and utero-vaginal prolapse in the group of female schizophrenics (Table 7.11). The deficit of menstrual disorders was particularly statistically

Table 7.9 Relative risks of diseases of gall-bladder and bone in cases of schizophrenia
and controls

Disease	No. cases	Schizophrenia		Affective psychosis		Depression	
		Observed No.	Relative Risk	Observed No.	Relative Risk	Observed No.	Relative Risk
Gall-bladder	1 771	5	0.37**	7	0.57	31	1.13
Osteomyelitis and other bone diseases	6 454	6	0.30***	7	0.60	37	1.17

** indicates probability $p < 0.01$.
*** indicates probability $p < 0.001$.

Table 7.10 Relative risks of certain urinary diseases and head injury in male schizophrenics and controls

Disease	No. cases	Schizophrenia		Affective psychosis		Depression	
		Observed No.	Relative Risk	Observed No.	Relative Risk	Observed No.	Relative Risk
Urinary calculus and other urinary diseases	7 521	1	0.11***	8	1.62	14	1.21
Head injury	13 711	31	1.4	15	2.00*	61	2.61***

  * indicates probability $p < 0.05$
*** indicates probability $p < 0.001$

Table 7.11 Relative risk of certain diseases in female schizophrenics
and controls

Disease	No. cases	Schizophrenia		Affective psychosis		Depression	
		Observed No.	Relative Risk	Observed No.	Relative Risk	Observed No.	Relative Risk
Appendicitis	7 586	1	0.14**	3	0.58	19	1.07
Varicose veins	3 884	2	0.18***	6	0.64	20	0.78
Menstrual disorders	7 416	11	0.40***	28	1.37	140	2.19***
Utero-vaginal prolapse	3 518	7	0.50*	13	0.89	35	1.10
Benign uterine fibroma	3 020	3	0.24**	13	1.27	29	1.00

   * indicates probability $p < 0.05$
  ** indicates probability $p < 0.01$
 *** indicates probability $p < 0.001$

significant. These conditions occurred as frequently as expected in females with affective psychoses. Depression was significantly associated with menstrual disorders.

Certain uncommon diseases

A number of relatively rare diseases have been postulated to be positively or negatively associated with schizophrenia, mainly in relation to the various biochemical aetiologies currently under consideration. The degree of statistical association was ascertained for several of these from the Oxford Record Linkage Study data (Table 7.12). In most instances the number of cases was very small and no significant associations were found.

Discussion

Results from the preliminary study of Oxford Record Linkage Study data, though subject to certain limitations, may help to clarify and focus attention on diseases apparently associated with schizophrenia which require more definitive study, and indicate the conditions which will have to be fulfilled for confident conclusions to be drawn. Three groups of disease associations can be considered: (1) common diseases previously said to be positively or negatively associated with schizophrenia but which have no generally accepted bearing on current aetiological theories. This group includes cancer, heart and vascular diseases, respiratory diseases and petic ucler; (2) diseases predicted from current aetiological hypotheses to be associated with schizophrenia including epilepsy, Parkinson's disease, rheumatoid arthritis, thyroid diseases, and a group of rare conditions and; (3) a group of disorders found to

Table 7.12 Observed and expected numbers of certain uncommon diseases in cases of schizophrenia

Disease	No. cases	Observed	Expected	Relative risk
Adrenal insufficiency	69	0	0.2	—
Anklosing spondylitis	118	0	0.43	—
Blindness	162	1	0.4	2.36
Chronic hepatitis	343	2	0.95	2.11
Hepato-lenticular degeneration	1 090	2	2.5	0.81
Hypopituitarism	119	0	0.4	—
Multiple sclerosis	492	1	1.8	0.57
Myasthenia gravis	39	0	1.1	—
Pernicious anaemia	591	2	1.7	1.15
Porphyria	94	1	0.3	3.55
Psoriasis	362	1	1.2	0.86
Systemic lupus erythematosus	516	1	1.6	0.64
Ulcerative colitis	382	0	1.4	—

be associated with schizophrenia but not previously reported and including gallbladder disease, osteomyelitis and other bone disease, certain urinary diseases, head injury, benign neoplasms, appendicitis, varicose veins, menstrual and uterine disorders.

Common diseases with no current aetiological implication

Cancer

The short period at risk covered by the data precludes definite conclusions about this important group of diseases. Although no reduction of relative risks for common cancers could be demonstrated, since duration of hospitalization was not taken into account, any effect comparable with that found by Katz *et al*. (1967) may have been masked. The small but significant excess of cancer of stomach is of interest in view of the similar finding of Opsahl (1933). The small excess of cancer of the breast in hebephrenic women may be a consequence of phenothiazine treatment (Ettigi *et al*. 1973). There is no doubt that much larger numbers of cases followed for a considerably longer time period are necessary to elucidate the relationship between schizophrenia and its sub-types with specific cancers. Duration of hospitalization must be taken into account in order to clarify any effects of the hospital environment and if data relating to the period over which antipsychotic drug therapy has been common practice are used, it may be impossible to control for its effects.

Heart and vascular disease

The highly significant excess of arteriosclerotic heart disease in both schizophrenics and controls is in accord with previous work by such authors as Alström (1942) and Ødegaard (1967), though the absence of excesses of other groups of vascular disorders in schizophrenics is of interest. Recently, Born (1979) has suggested that chlorpromazine may reduce the incidence of myocardial infarction, citing the observation of Ødegaard (1967) that the incidence of cardiac disease had not increased over community rates in schizophrenics to the same extent as in other mental patients. Hussar (1962) found no increase in mortality from acute myocardial infarction since tranquillizers had been introduced though there had been an increase in arteriosclerotic heart disease.

Respiratory disease

Increased mortality and morbidity from pneumonia and other respiratory and infectious diseases among mental-hospital patients has been widely reported though the excess seems to have been less in more recent times. It is likely that the environmental conditions in mental hospitals coupled with relatively poor hygiene and behavioural anomalies associated with some mental illness are conducive to the spread of such infectious diseases, especially among the aged and infirm. Schizophrenics were formerly subject to marked excess mortality from pulmonary tuberculosis though this has virtually disappeared following introduction of specific treatment and improvement in the physical condition of the patients. It is therefore of some interest that there was an increased relative risk of pulmonary tuberculosis prior to the diagnosis of schizophrenia in the Oxford data.

Peptic ulcer

No association between gastric and duodenal ulceration and schizophrenia was found in the Oxford data, though increased incidence has been reported previously (Ehrentheil 1957) and excess mortality from this cause was found among Filipino males in Hawaii (Weiner and Marvit 1977).

Diseases predicted form current aetiological hypotheses

Epilepsy

It is surprising that the Oxford data showed no relation between schizophrenia and minor or focal epilepsy in view of the work of Slater and Beard (1963) and the more recent finding of excess schizophrenia in temporal lobe epilepsy by Taylor (1975). Deficient ascertainment of minor and focal epilepsy in the Oxford data seems likely. The high relative risks of major epilepsy, epileptic psychosis, and unspecified epilepsy among paranoid and non-specific schizophrenias are of interest but do not appear to lend support to the biochemical explanation of a negative association offered by Reynolds (1968) and later elaborated by Levi and Waxman (1975). Results from the Oxford data indicate a complex relationship between schizophrenia and epilepsy which requires detailed study.

Parkinson's disease

Parkinsonism is a common side-effect of phenothiazine therapy so that an increased relative risk following the diagnosis of schizophrenia is not surprising. The occurrence of a schizophrenia-like episode of psychosis in the course of long-standing Parkinsonism has been described by Crow *et al.* (1976) who suggested it might be relevant to the 'dopamine hypothesis' of the aetiology of schizophrenia.

Allergic conditions

Reduced incidence of asthma, hay fever, and other allergic conditions have been reported, as have alternating psychotic and allergic states. Despite a number of studies, the role of histamine in schizophrenia is still unclear (Matthysse and Lipinski 1975). Although the low relative risk obtained in the Oxford data was not significant, there was a contrast with the high relative risks found in the control groups. It seems possible that there may be a low incidence, at least in one or more of the sub-types of schizophrenia, and further detailed study is indicated.

Diabetes mellitus

'Antagonism' between schizophrenia and diabetes mellitus has been postulated by Rassidakis *et al*. (1974). There was no support for this hypothesis in the Oxford data.

Malabsorption

The marginally significant increase in relative risk of malabsorption in schizophrenics obtained when one simultaneously diagnosed case was added to the original analysis must be interpreted with caution since the group may include several forms of malabsorption and confirmation of the diagnoses would be necessary to validate the result. The single case of hebephrenia compared with 0.014 expected may have been a chance occurrence and little weight can be attached to this finding. Apart from the therapeutic studies of Dohan and his colleagues (Dohan *et al*. 1969) and the gluten challenge test of Singh and Kay (1976), several studies of various kinds have not confirmed the suspected association with coeliac disease (Dean *et al*. 1975; Stevens *et al*. 1977). Though the Oxford results do not rule out the possibility, it seems doubtful that the hypothesis of gluten sensitivity can be sustained.

Arthritis

The reduction in relative risk of rheumatoid arthritis in schizophrenia is in accord with a number of earlier studies and no study has contradicted this negative association (Baldwin 1979). Nevertheless, the possibility of low ascertainment cannot be dismissed since this condition may be treated without transferring patients from psychiatric hospitals to other units and in these instances there would be no record of the arthritic condition in the Oxford Record Linkage Study. Against this must be set the finding of normal risks for this disease in both control groups, suggesting that under-reporting may not be an important factor in the estimated relative risk in schizophrenia.

The recent extensive Swedish study by Österberg (1978) considered the aetiological implications of a reduced risk of rheumatoid arthritis in schizophrenia and other psychiatric disorders. If the risk was reduced in psychiatric illness generally, a genetic or biochemical explanation would be unlikely and personality characteristics would be more likely to account for the relationship. The Oxford data indicate that rheumatoid arthritis is not unusually uncommon in affective psychosis and non-psychotic depression, the evidence being that the reduced risk in schizophrenia is specific. Thus it is reasonable to seek explanations in genetic and biochemical terms.

The only current biochemical hypothesis which predicts a low relative risk of rheumatoid arthritis in schizophrenia is that of prostaglandin deficiency (Horrobin *et al*. 1978). This postulates that rheumatoid arthritis should be uncommon except in association with schizo-affective and catatonic states. Several of Österberg's (1978) cases were said to be schizo-affective, but the majority, like both the cases in the Oxford study, were non-specific schizophrenia.

Österberg (1978) also discussed the aetiological consequences of relationships between schizophrenia and various other joint diseases, particularly those known to be associated with human leucocyte antigens (HLA). Her data seemed to indicate that ankylosing spondylitis might be over-represented in schizophrenics in comparison with rheumatoid arthritis. The data in the present Oxford study were insufficient to test this suggestion, with no cases found and none expected. This was also the position with respect to other possibly relevant diseases, including systematic lupus erythematosus and psoriasis. The only other result of interest in this group of conditions was a marginally significant reduction of relative risk of osteoarthritis in schizophrenics, especially in males, which did not occur in the control groups.

The reduced relative risk of rheumatoid arthritis in schizophrenia, though not indisputably confirmed, seems likely to be real and specific. This negative association presents one of the most promising leads for further aetiological research and clearly requires further study both in itself and in relation to other joint diseases and their HLA associations. Although some studies of HLA associations with schizophrenia have been carried out, the results so far are inconclusive (McGuffin *et al*. 1978). The D locus particularly, which is thought to be associated with rheumatoid arthritis (Stastny 1976), has not been studied in schizophrenics. Nevertheless, the contrast between the low relative risk of rheumatoid arthritis and the purported raised relative risk of coeliac disease, which is also associated with the D antigen, perhaps suggests that HLA associations with schizophrenia, if they exist, are not straightforward. Indeed, it may be argued that the search for a genetic marker through the HLA system is premature and that a wide-ranging search among as many known markers as possible would be a more logical preliminary step (Böök *et al*. 1978).

In the light of these observations, the Oxford data were analysed to estimate the degree of association between rheumatoid arthritis and malabsorption and yielded a relative risk of 4.68 ($p = 0.01$). Coeliac disease has been shown to be associated with focal epilepsy (Chapman *et al*. 1978) and this relationship was also found in the Oxford data (observed = 1; expected = 0.2; $p = 0.02$). The interrelationship between these diseases was further demonstrated by the finding of a highly significant association between focal epilepsy and rheumatoid arthritis (relative risk = 20.13; $p = 0.0005$). Though the number of cases in each of these associations was small, the strength of the statistical relationship is such as to suggest a possibly useful lead for further study.

Thyroid disorders

Both thyrotoxicosis and myxoedema were predicted to be associated with schizophrenia by Gilka (1975) in an extensive review of biochemical studies. Horrobin (1979) noted that thyroid deficiency may reduce prostaglandin secretion. If this were the case, myxoedema should be more common than expected in schizophrenia. There was no evidence of an increased risk of hyperthyroidism in the Oxford data, but there were significant excesses of myxoedema in catatonics and non-specific schizophrenics, and in both control groups.

Rare diseases

Both the hypothesis advanced by Gilka (1975), of one or more metabolic defects in the tryptophan–niacin pathway, and the prostaglandin-deficiency hypothesis of Horrobin (1979), predict associations between schizophrenia and a number of relatively uncommon diseases. The difficulty with virtually all of these is their rarity, since to obtain a robust estimate of relative risks, a very large sample of schizophrenics would have to be obtained and related to a defined population at risk. The observed and expected numbers shown in Table 7.12 can be used to infer the size of sample which might be required. Much more extensive data should be available from the Oxford Record Linkage Study in the near future and may enable estimation of some of these associations.

Previously unreported associations

Significant reductions of relative risk were obtained for a number of diseases for which no previous reports have been found. Some of these were sex specific, and comparable reductions of risk were not found among the control groups. Although some of these results might be presumed to stem from known characteristics of schizophrenics, such as the reduced fertility of schizophrenic women which might account for the low incidence of utero-vaginal prolapse, and the reputed high tolerance of pain and discomfort, which might reduce the likelihood of discovery of some disorders, it seems preferable to presume more prosaic explanations until further research can be undertaken. Deficient ascertainment due to the limited span of the available data, or caused by treatment of some conditions within the psychiatric hospital are at least as likely as failure of clinical detection through infrequent or inadequate physical examination or lack of patient reports. Yet not all this group of negative associations can be accounted for in these ways. For example, appendicitis is virtually certain to be detected and could not have been treated in the psychiatric hospitals concerned. Thus there is reason to suppose that some at least of these reduced risks are real rather than artefactual, so that further study should be contemplated.

The only condition in this group with a high relative risk was head injury in male schizophrenics, which was more common than expected after discharge with schizophrenia.

Conclusions

This preliminary study of data from the Oxford Record Linkage Study on physical morbidity in schizophrenia has been useful in clarifying some of the large number of previously reported positive and negative disease associations. By using epidemiological methods to obtain comparable relative risks in both schizophrenia and control groups, it has focused attention on a group of diseases which warrant further study and has indicated directions for future research likely to be relevant to aetiological understanding of schizophrenia.

With a few exceptions, most of the diseases predicted to be associated with schizophrenia from current aetiological hypotheses are so uncommon that valid estimates of relative risk cannot be obtained until much larger collections of data become available. The exceptions which are most likely to repay further study are epilepsy, Parkinson's disease, and rheumatoid arthritis. It is fairly well established that there is a reduced incidence of rheumatoid arthritis in schizophrenia but not in some other psychiatric disorders and this negative association perhaps holds the most promise of illuminating schizophrenia aetiology.

References

ACHESON, E. D. (1967). *Medical Record Linkage*. Oxford University Press.

ALSTRÖM, C. H. (1942). Mortality in mental hospitals with special regard to tuberculosis. *Acta Psychiatr. Neurol.* **24** (Suppl.).

BALDWIN, J. A. (1973). Linked record medical information systems. *Proc. R. Soc. (Lond.)* **184**, 403.

—— (1979). Schizophrenia and physical disease: a review. *Psychol. Med.* **9**, 611.

BÖÖK, J. A., WETTERBERG, L., AND MODRZEWSKA, K. (1978). Schizophrenia in a North Swedish geographical isolate, 1900–1977; epidemiology, genetics, and biochemistry. *Clin. Genet.* **14**, 373.

BORN, G. V. R. (1979). Possible role for chlorpromazine in protection against myocardial infarction. *Lancet* **1**, 822.

CHAPMAN, R. W. G., LAIDLOW, J. M., COLIN-JONES, D., EADE, O. E., AND SMITH, C. L. (1978). Increased prevalence of epilepsy in coeliac disease. *Lancet* **2**, 250.

CROW, T., JOHNSTONE, E. C., AND MCCLELLAND, H. A. (1976). The coincidence of schizophrenia and Parkinsonism: some neurochemical implications. *Psychol. Med.* **6**, 227.

DEAN, G., HANNIFFY, L., STEVENS, F. M., TEMPERLEY, L., O'BROIN, J. D., SCOTT, J., AND CAHALANE, S. F. (1975). Schizophrenia and coeliac disease. *J. Ir. Med. Assoc.* **68**, 545.

DOHAN, F. C., GRASBERGER, F. M., LOWELL, H. T., JOHNSTON, A., AND ARBEGAST, W. (1969). Relapsed schizophrenics: more rapid improvement on a milk- and cereal-free diet. *Br. J. Psychiatry* **115**, 595.

EHRENTHEIL, O. F. (1957). Common medical disorders rarely found in psychotic patients. *Arch. Neurol. Psychiatr.* **77**, 178.

ETTIGI, P., LAL, S., AND FRIESEN, H. G. (1973). Prolactin, phenothiozines, admission to mental hospital and carcinoma of the breast. *Lancet* **2**, 266.

GILKA, L. (1975). Schizophrenia: a disorder of tryptophan metabolism. *Acta Psychiatrica Scand.* **258** (Suppl.).

HORROBIN, D. F. (1979). Schizophrenia: reconciliation of the dopamine, prostaglandin, and opioid concepts and the role of the pineal. *Lancet* **1**, 529.

——, ALLY, A. I., KARMALI, M., KARMAZYN, M., MANKU, M. S., AND MORGAN, R. O. (1978). Prostaglandins and schizophrenia: further discussion of the evidence. *Psychol. Med.* **8**, 43.

HUSSAR, A. E. (1962). Effect of tranquillizers on medical morbidity and mortality in a mental hospital. *J. Am. Med. Assoc.* **179**, 682.

KATZ, J., KUNOESKY, S., PATTEN, R. E., AND ALLAWAY, N. C. (1967). Cancer mortality among patients in New York mental hospitals. *Cancer* **20**, 2194.

LEVI, R. N. AND WAXMAN, S. (1975). Schizophrenia, epilepsy, cancer, methionine, and folate metabolism. *Lancet* **2**, 11.

MATTHYSSE, S. AND LIPINSKI, J. (1975). Biochemical aspects of schizophrenia. *Annu. Rev. Med.* **26**, 551.

McGUFFIN, P., FARMER, A. E., AND RAJAH, S. M. (1978). Histocompatibility antigens and schizophrenia. *Br. J. Psychiatry* **132**, 149.

ØDEGAARD, Ø. (1967). Mortality in Norwegian psychiatric hospitals, 1950–1962. *Acta Genetica* **17**, 137.

OPSAHL, R. (1933). Frequency of cancer among patients with mental diseases. *Norsk Magasin for Laegevidenskapen* **94**, 771.

ÖSTERBERG, E. (1978). Schizophrenia and rheumatic disease. *Acta Psychiatr. Scand.* **58**, 339.

RASSIDAKIS, N. C., EROKOCRITOU, ATHITAKIS, KARAYANNIS, AND KARAIOSSE-PHIDES, K. (1974). Schizophrenia, psychosomatic illnesses, diabetes mellitus, and malignant neoplasms. *Int. Ment. Health Res. Newsletter* **21**, 12.

REYNOLDS, E. H. (1968). Epilepsy and schizophrenia: relationship and biochemistry. *Lancet* **1**, 398.

SINGH, M. M. AND KAY, S. R. (1976). Wheat gluten as a pathogenic factor in schizophrenia. *Science* **191**, 401.

SLATER, E. AND BEARD, A. W. (1963). The schizophrenia-like psychoses of epilepsy: (i) psychiatric aspects. *Br. J. Psychiatry* **109**, 95.

STASTNY, P. (1976). Mixed lymphocyte cultures in rheumatoid arthritis. *J. Clin. Invest.* **57**, 1148.

STEVENS, F. M., LLOYD, R. S., GERAGHTY, S. M. J., REYNOLDS, M. T. G., SARSFIELD, M. J., McNICHOLL, B., FOTTRELL, P. F., WRIGHT, R., AND McCARTHY, C. F. (1977). Schizophrenia and coeliac disease—the nature of the relationship. *Psychol. Med.* **7**, 259.

TAYLOR, D. C. (1975). Factors influencing the occurrence of schizophenia-like psychosis in patients with temporal lobe epilepsy. *Psychol. Med.* **5**, 249.

WEINER, B. P. AND MARVIT, R. C. (1977). Schizophrenia in Hawaii: analysis of cohort mortality risk in a multi-ethnic population. *Br. J. Psychiatry* **131**, 497.

8

Hospital morbidity in early life in relation to certain maternal and foetal characteristics and events at delivery

E. D. ACHESON

Introduction

Using a system of linked medical records (Acheson 1964) it is possible not only to accumulate and to correlate information about successive events as they occur to individuals, but to bring together information about different members of a family (Newcombe, *et al*. 1959; Newcombe 1964). This study exemplifies a simple application of this approach in which information about two classes of individuals related by blood (i.e. mothers and their babies) has been brought together from five separate categories of medical record for a defined population at risk.

An analysis has been made of the in-patient hospital morbidity experienced during 1962 of 5901 babies born in that year to mothers resident at the time of delivery in the City of Oxford, Oxfordshire (except Henley Municipal Borough and Rural District Council), and Abingdon Municipal Borough and Rural District in Berkshire. This excludes 147 babies who were stillborn or who died before discharge from the maternity hospital in which they were born, and 379 babies who were born to resident mothers in hospitals outside the area for whom information about the relevant pregnancy and confinement were lacking. For the remainder, information was brought together from the following sources:

Baby: Birth certificate.
           Hospital discharge summaries (if any).
           Death certificate (if any).
Mother: Domiciliary confinement record (for mothers delivered at home)
           or
           Hospital confinement record (for mothers delivered in hospital).

The records were checked with a master index to bring together different records relating to the same person, the birth certificate being used as a 'bridge record' to collate the records of mother and baby.

Reprinted from *British Journal of Preventive and Social Medicine* **19**, 164–73 (1965).

A summary punched card was then prepared relating to events in the calendar year 1962 only. The present chapter deals with some simple analyses of the in-patient morbidity of the babies in relation to maternal characteristics and selected events at delivery:

$$\text{IN-PATIENT MORBIDITY} = \frac{\text{(No. of babies discharged from hospital once or more in 1962)}}{\text{(Total no. of babies born alive who were living on discharge from the maternity hospital)}} \times 100$$

It should be stressed that as the period of study was a calendar year the average period of exposure of the babies to the risk of admission to hospital is approximately 6 months, babies born on 1 January 1962, having a whole year's exposure, while at the other extreme for babies born on 31 December 1962, the exposure was nil. When the data of subsequent years are to hand, it will be possible to take into account age at admission to hospital, and to consider separately, for example, the morbidity of the first month and of the remainder of the first year.

Results

General

Diagnosis reached during first spell of in-patient care

203 (3.4 per cent) of the 5901 babies were discharged from hospital in the calendar year 1962 once or more. Of these, 34 babies (16.7 per cent) had been readmitted or transferred at least once during the calendar year, the accumulated number of in-patient spells of treatment being 255. In Table 8.1, the 203 babies are distributed by principal diagnoses reached during the first hospital spell, grouped according to the major divisions of the International Statistical Classification of Diseases, Injuries, and Causes of Death (WHO 1957).

As was expected, the largest single cause of admission (29.0 per cent) was a congenital malformation. The other large groups were digestive and genito-urinary diseases (14.8 per cent), diseases peculiar to infancy, predominantly feeding difficulties (13.8 per cent), and respiratory diseases, almost exclusively infections (13.3 per cent).

Place of delivery

Table 8.2 shows that the in-patient morbidity of babies delivered at home (3.2 per cent) was almost identical to that of those delivered in hospital (3.5 per cent). This is surprising in view of the selection in this area of difficult cases and of nulliparae and grand multiparae for delivery in hospital

Table 8.1 Principal diagnosis of first spell in hospital for 203 babies born in 1962 who were subsequently discharged from hospital at least once in the same year

Principal diagnosis of first spell in hospital*		All Deliveries	
Code	Disease	No.	%
001–158	Infective and parasitic	7	3.4
140–239	(except 228) Neoplasms	0	0.0
240–289	Allergic, endocrine, metabolic, nutritional	2	1.0
290–299	Blood diseases	0	0.0
300–326	Mental diseases (except 325)	0	0.0
330–397	Nervous system and special senses (except 389)	6	3.0
400–468	Circulatory	4	2.0
470–527	Respiratory	27	13.3
530–637	Digestive and genito-urinary (except 560, 561)	30	14.8
690–716	Skin	2	1.0
720–749	Bones, etc. (except 748)	2	1.0
750–759	Congenital malformations 228, 325, 351, 389, 398, 560, 561, 748	59	29.0
760–776	Infancy	28	13.8
780–789	Symptoms, etc.	16	7.9
800–999	Accidents, etc.	13	6.4
	Others	7	3.4
Total		203	100

* Bracketed figures in this Table and Table 8.3 refer to the diagnostic rubrics of the International Classification of Diseases, Injuries and Causes of Death (7th revision).

Table 8.2 In-patient morbidity of the baby in 1962 by place of delivery

Place of delivery		Babies admitted to hospital		Total no. of babies
		No.	%	
Consultant	A	62	3.7	1684
Hospital	B	36	3.3	1094
	*C	22	3.5	623
	All	120	3.5	3401
GP unit	A	5	1.9	264
	B	24	6.4	372
	C	2	0.9	233
	All	31	3.6	869
All hospitals		151	3.5	4270
At home		52	3.2	1631
Total		203	3.4	5901

* A mixed consultant and GP unit.

(Acheson and Evans 1964). When all consultant obstetric units are compared with all GP obstetric units, the subsequent morbidity of the baby in the two groups is also closely similar. However, within general practitioner units as a group there is substantial variation. Thus for babies born at GP unit B the subsequent morbidity was 6.4 per cent, while for GP units A and C the figures are 1.9 per cent and 0.9 per cent respectively. It seems likely that GP unit B's high figure is due in part at least to the presence in the same town of hospital beds for children under the control of GPs.

Although the numbers are small, the in-patient morbidity in four diagnostic groups has been compared for hospital and home deliveries in Table 8.3. The discharge rate for congenital malformations, and for nutritional maladjustment, etc., was similar in the two groups, but was slightly higher for infections in babies delivered in hospital than in those delivered at home. This was compensated for by a higher rate for miscellaneous causes among babies delivered at home than among those delivered in hospital.

Plurality

136 babies of multiple births survived delivery and were discharged alive. In this group, as might be expected, the subsequent hospital morbidity was higher (6.6 per cent) than in singletons (3.4 per cent), but the difference is not significant.

Table 8.3 Babies discharged from hospital once or more and in-patient morbidity (%). In four diagnostic groups, by place of delivery

Diagnostic group	Place of delivery			
	Hospital		Home	
	No.	In-patient morbidity (%)	No.	In-patient morbidity (%)
Congenital malformations (750–759, 228, 325, 351, 389 398, 560, 561, 748)	44	1.0	15	0.9
Infections †	41	0.9	9	0.6
Nutritional maladjustment and symtoms relating to the alimentary tract ‡ (772, 784, 785)	16	0.4	4	0.2
Trauma (800–999)	8	0.2	5	0.3
All Other	42	1.0	19	1.2
Total	151	3.5	52	3.2

† 001–138, 470–502, 571, 763–768, and miscellaneous infections elsewhere.

‡ Excludes diarrhoea in infancy.

Maternal factors

Maternal age

The in-patient morbidity of the babies is given by maternal age at time of delivery in Table 8.4 and Fig. 8.1. It is highest (6.2 per cent) in babies of mothers under age 20, after which it falls, reaching a minimum of 2.8 per cent in the 25- to 29-year age group, with a slight increase thereafter. When illegitimate births were excluded, the excess morbidity in the babies of young mothers remained unaltered (Fig. 8.1B), but the secondary rise in the babies of mothers over age 30 seen in Fig. 8.1A disappeared. The difference in morbidity between the babies of mothers in the youngest age group and the remainder is significant ($t = 2.7$; $P < 0.01$). The findings were similar when babies delivered at home and in hospital were considered separately.

Figure 8.1C shows the effect of maternal age in nulliparae and multiparae separately and demonstrates that babies of young mothers have a high hospital morbidity in each of these groups. A similar analysis, taking into account social class of mother, shows that when Classes I, II, and III and

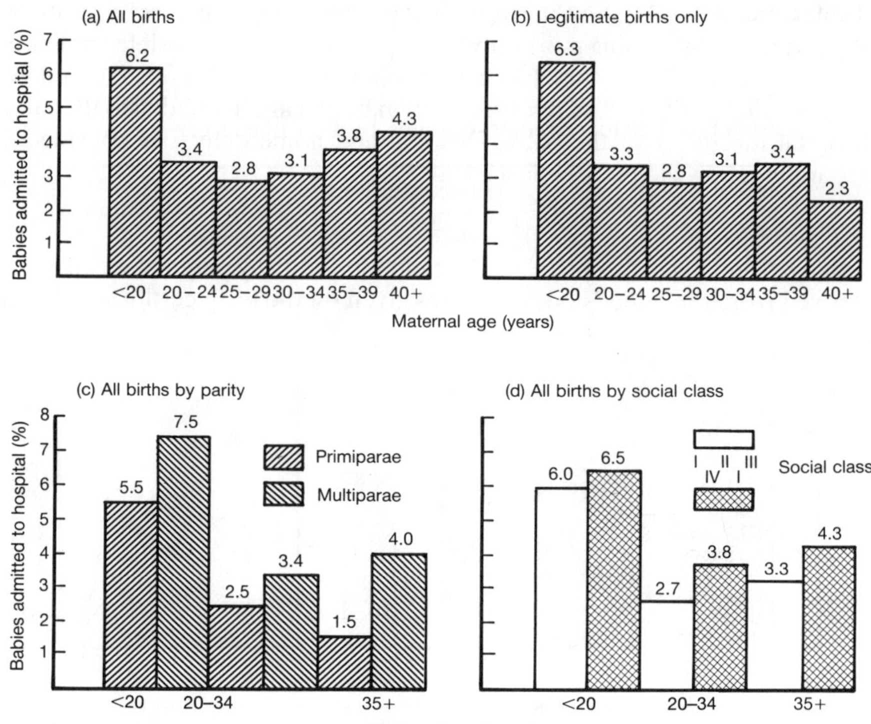

Fig. 8.1 In-patient morbidity of the baby in 1962, according to maternal age.

Table 8.4 In-patient morbidity of the baby in 1962 by maternal age

Maternal age at time of delivery (yrs.)	Babies discharged from hospital		Total no. of babies
	No.	%	
19 or under	30	6.2	487
20–24	65	3.4	1928
25–29	49	2.8	1739
30–34	33	3.1	1078
35 or over	26	3.9	663
Not known	—	—	6
Total	203	3.4	5901

Classes IV and V are grouped (the armed forces, students, and unclassified cases were excluded), the hospital morbidity in each is highest in young mothers (Fig. 8.1D).

Parity

Table 8.5 and Fig. 8.2 show the in-patient morbidity of the babies by parity. A slight trend of increasing morbidity with increasing parity is visible but this is not significant. The result, taken in conjunction with Figs. 8.1A and B, suggests that parity is less important than maternal age in relation to the early hospital morbidity of the baby. Hospital and home deliveries considered separately show almost identical trends.

Social class

Table 8.6 and Fig. 8.2 depict the in-patient morbidity by parental social class, derived from the father's occupation as given on the birth certificate. When

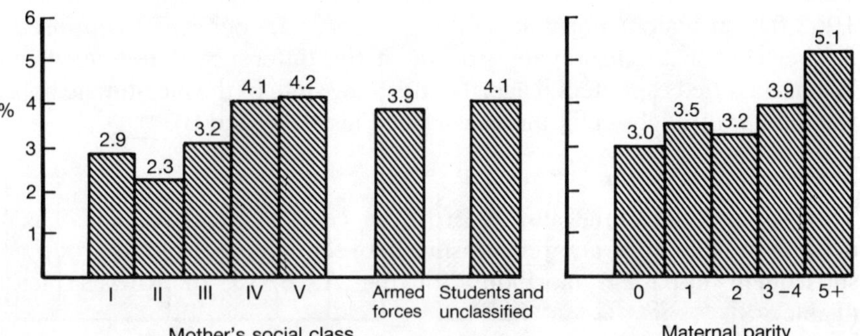

Fig. 8.2 In-patient morbidity of the baby in 1962, according to parental social class and parity.

Table 8.5 In-patient morbidity of the baby in 1962 by maternal parity

Maternal parity	Babies discharged from hospital		Total no. of babies
	No.	%	
Nulliparae	62	3.0	2088
1	62	3.5	1753
2	34	3.2	1058
3 and 4	28	3.9	709
5 or over	14	5.1	274
Not known	3	0.0	19
Total	203	3.4	5901

all cases are considered together, an increasing trend of morbidity with diminishing social class is seen, which does not quite reach the conventional limit of statistical significance. (When armed forces, students, and unclassified cases are excluded, $\chi^2 = 5.5$; d.f. $= 2$, $P < 0.1 > 0.05$.) This trend is slightly more marked for home deliveries considered alone ($\chi^2 = 6.6$; d.f. $= 2$; $P < 0.05 > 0.01$), and less marked for hospital deliveries alone.

Maternal past history of stillbirths or miscarriage

Table 8.7 shows that a past history of maternal miscarriage or stillbirth did not appear to influence the in-patient morbidity of the baby in the early months. The relationship of these factors was also examined separately in babies of mothers under or over 20 years of age, but no trends were seen.

Legitimacy

256 babies who survived delivery and were discharged from hospital alive were known to be illegitimate. This almost certainly underestimates the true state of affairs as our means of ascertainment of legitimacy was inadequate in 1962. The in-patient morbidity in this group was 5.9 per cent, as opposed to 3.3 per cent in the legitimate group, but the difference is not significant. When classified by maternal age, the morbidity rate of the illegitimate babies was similar throughout the age range (see Figs. 8.1A and B).

Maternal blood group

This was recorded in all but 393 (6.7 per cent) of the cases. There is no definite indication of any relationship between maternal blood group and subsequent in-patient morbidity for the A, B, O, or Rhesus factors (Table 8.8).

Table 8.6 Effect of maternal social class on in-patient morbidity of the baby in 1962

Maternal Social class	All cases		Total no. of babies	Home deliveries		Total no. of babies	Hospital deliveries		Total no. of babies
	Babies discharged from hospital			Babies discharged from hospital			Babies discharged from hospital		
	No.	%		No.	%		No.	%	
Class I	13	2.9	450	3	1.3	232	23	2.9	782
Class II	13	2.3	564						
Class III	80	3.2	2466	17	2.5	685	63	3.5	1781
Class IV	43	4.1	1047	23	4.4	520	46	4.0	1141
Class V	26	4.2	614						
Armed forces	16	3.9	408						
Students	0	0.0	63	9	4.6	194	19	3.4	566
Unclassified	12	4.1	289						
Total	203	3.4	5901	52	3.2	1631	151	3.5	4270

Table 8.7 Effect of a maternal past history of stillbirth or miscarriage on in-patient morbidity of the baby in 1962

Maternal history of stillbirth or miscarriage	Babies discharged		Total no. of babies
	No.	%	
No stillbirths or miscarriages	165	3.3	4981
1, 2, 3, or more stillbirths	4	3.0	135
1, 2, 3, or more miscarriages	32	4.4	729
1 or more stillbirths and 1 or more miscarriages	0	0.0	20
Not known or not stated	2	5.9	34
Total	203	3.4	5899

Obstetric factors and factors relating to the baby

Birthweight

The influence of birthweight on the in-patient morbidity is shown in Table 8.9 and Fig. 8.3. As might be expected, babies weighing less than 5.5 lb. (2490 g) who survive the perinatal period have a significantly higher in-patient morbidity than the remainder ($t = 2.4$; $P < 0.05$). For weights higher than 5.5 lb. the in-patient morbidity declines to a minimum of 2.3 per cent in the 8–9 lb. group with a slight rise thereafter. It is interesting that the lowest morbidity figure should be for babies with weights higher than both

Table 8.8 In-patient morbidity of the baby in 1962 by maternal blood group

Maternal blood group	Babies discharged from hospital		Total no. of babies
	No.	%	
A	80	3.3	2442
AB	2	1.3	155
O	91	3.7	2441
B	15	3.2	470
Not known	15	3.8	393
Total	203	3.4	5901
Rhesus +	153	3.4	4503
Rhesus −	35	3.5	1005
Not known	15	3.8	393
Total	203	3.4	5901

Table 8.9 Effect of birth weight on in-patient morbidity
of the baby in 1962

Baby's Birth weight (lb)	Babies discharged from hospital		Total no. of babies
	No.	%	
4.5	8	9.2	87
5.5	13	6.0	217
6.5	39	4.5	866
7.5	66	3.3	1975
8	34	3.3	1041
9	29	2.3	1249
10	9	2.5	354
10 or more	4	4.4	90
Not weighed	1	4.5	22
Total	203	3.4	5901

the median (6.9 lb.) and the average (7.5 lb.). A similar trend is shown for perinatal mortality taken from the same material (ORLS area, 1962). It seems therefore that the average birthweight may not be the optimal (Karn and Penrose 1951).

Type of delivery

This was classified according to the International Classification into the following groups:
(1) Spontaneous: (*a*) uncomplicated; (*b*) complicated.
(2) Manipulation without instruments.
(3) Forceps delivery (all types).
(4) Caesarean section.

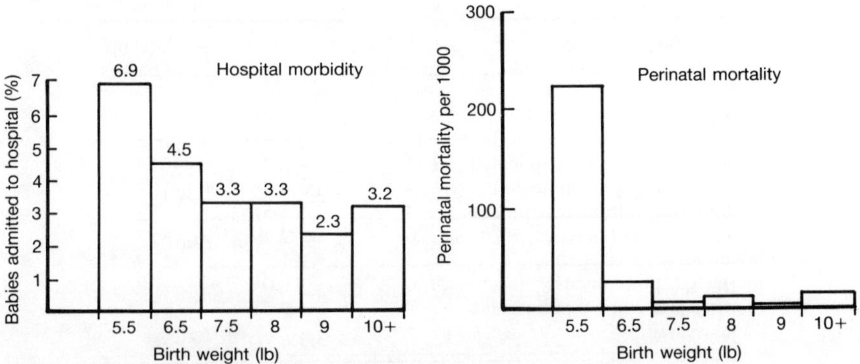

Fig. 8.3 In-patient morbidity and perinatal mortality in 1962, by birthweight.

(5) Unassisted.
(6) All other (including vacuum delivery).

Uncomplicated spontaneous deliveries are followed by a hospital morbidity (3.3 per cent) very close to the average (Table 8.10). Of the remainder, the small group of unassisted spontaneous deliveries has the highest morbidity. It is interesting that after forceps delivery and Caesarean section the morbidity is less than average. Further analysis, taking into account birthweight, maternal age, and type of morbidity, will be undertaken when further material is available.

Length of second stage of labour

In 1962 an attempt was made to determine the length of the second stage of labour from the obstetric records, as it was felt that this might provide an index of foetal stress. Table 8.11 shows that, when all cases (excluding 173 Caesarean sections) are considered, there is no clear evidence of any relationship between the two variables. Removal of small babies (weighing less than 5.5 lb.) from the analysis did not affect the issue. The second stage of labour is notably difficult to define and to measure, and it is possible that the negative result is an indication of this.

Maternal anaesthesia

In Table 8.12 the babies are classified according to whether an anaesthetic was administered to the mother for the delivery; the type of anaesthesia is also given. The results are negative. It is particularly interesting that the substantial number of babies whose mothers had a general or spinal anaesthetic have a similar hospital morbidity experience to those whose mothers had no anaesthetic.

Table 8.10 In-patient morbidity of the baby in 1962 by type of delivery

Type of delivery	Babies discharged from hospital		Total no. of babies
	No.	%	
Spontaneous (a) uncomplicated	134	3.3	4017
(b) complicated	34	4.8	703
Manipulation without instruments	4	5.5	73
Forceps delivery (all types)	20	2.3	853
Caesarean section	3	1.7	174
Unassisted spontaneous	6	10.0	60
All other (including vacuum) and unclassified	2	9.5	21
Total	203	3.4	5901

Table 8.11 In-patient morbidity of the baby in 1962 by duration
of second stage of labour

Length of second stage of labour (min.)	Babies discharged from hospital		Total no. of babies
	No.	%	
30	130	3.7	3538
60	37	3.4	1084
More than 60	16	2.6	622
Not applicable (Caesarean section or not known)	20	3.0	657
Total	203	3.4	5901

Administration of oxygen to the infant at birth

A record of the administration of oxygen to the infant was found in 381 cases
(6.4 per cent), of which almost all (370) were delivered in hospital. As the
administration of oxygen might be expected to correlate with the existence of
asphyxia at birth, it was interesting and unexpected to find that the subsequent morbidity was identical (3.4 per cent) whether or not oxygen was
given.

Baby's blood group

Blood derived from the umbilical cord had been typed in respect of the ABO
and Rhesus factors in 1361 babies who survived delivery and discharge from
hospital. The subsequent in-patient morbidity in each of the four ABO

Table 8.12 In-patient morbidity of the baby in 1962 by nature
of anaesthetic administered to mother at delivery*

Anaesthetic administered to mother at delivery	Babies discharged from hospital		Total no. of babies
	No.	%	
None	113	3.6	3124
General	8	2.6	306
Spinal	9	3.4	263
Local	2	2.5	79
Pudendal	10	2.5	405
Unknown	9	9.7	93
Total	151	3.5	4270

* Combination of local or pudendal not done.

Table 8.13 In-patient morbidity of the baby in 1962 by baby's blood group

Baby's blood group	Babies discharged from hospital		Total no. of babies
	No.	%	
A	19	3.1	608
AB	3	7.7	39
O	18	3.0	603
B	3	2.7	111
Unclassified	108	3.7	2909
Total	151	3.5	4270
Rhesus +	33	3.0	1111
Rhesus −	10	4.0	250
Unclassified	108	3.7	2909
Total	151	3.5	4270

phenotypes is shown in Table 8.13. No significant differences were demonstrated. The Rhesus positive and negative groups are given in Table 8.13; there was no material difference in morbidity between these two groups.

Factors associated with domiciliary deliveries

Type of assistance at delivery

A general practitioner was present at 430 (26.4 per cent) of the deliveries which took place at home. The subsequent morbidity of these babies (3.2 per cent) was identical to that of those babies at whose delivery a midwife was the senior person present. Babies whose mothers had no assistance at delivery had a higher subsequent morbidity (7.3 per cent) but the number was small and the difference from the remainder is not significant (Table 8.14).

Week in pregnancy of midwife's first antenatal visit

For mothers booked and delivered at home, a striking relationship was demonstrated between the week of gestation in which the first antenatal visit of the midwife took place and the subsequent hospital morbidity of the baby after delivery (Table 8.15 and Fig. 8.4). This trend is significant ($\chi^2 = 9.6$; d.f. = 2; $P < 0.01$). It was thought at first that the trend might be due to a bias introduced by the absence from the table of 143 mothers booked for confinement at home but later transferred for delivery in hospital. However, the subsequent morbidity of the babies of these mothers (not shown) was only slightly higher than average (4.2 per cent) and could not have affected the issue. It seems possible that the relationship between the date of the first antenatal visit and subsequent morbidity may indicate that mothers who seek

Table 8.14 In-patient morbidity of the baby in 1962 by nature of assistance at delivery (domiciliary)

Persons present at delivery	Babies discharged from hospital		Total no. of babies
	No.	%	
Midwife	34	3.0	1149
GP with or without midwife	24	3.2	430
Unassisted	3	7.7	39
Unknown	1	7.7	13
Total	52	3.2	1631

medical aid late in pregnancy are in general those who later have difficulty in caring for the baby.

Discussion

The index of the babies' morbidity experience used here is a crude and defective one because it relates only to that fraction of illness which is treated in hospital. The size of this fraction depends not only upon the nature and severity of the illness but upon the social circumstances of the family, the policy of the general practitioner, and the availability of paediatric hospital beds. The variety of disease covered is wide and ranges from the extreme instance of conditions which would compel admission in any circumstances (e.g. meningitis, pyloric stenosis, strangulated hernia, meningocele) to those

Table 8.15 In-patient morbidity of the baby in 1962 by date of midwife's first antenatal visit during mother's pregnancy

Week in pregnancy of midwifes first antenatal visit	Babies discharged from hospital		Total no. of babies
	No.	%	
19	10	1.4	706
20–24	12	3.3	360
25–29	8	3.2	251
30–34	13	6.6	196
35 +	7	8.5	82
No visit or not known	2	5.6	36
Total	52	3.2	1631

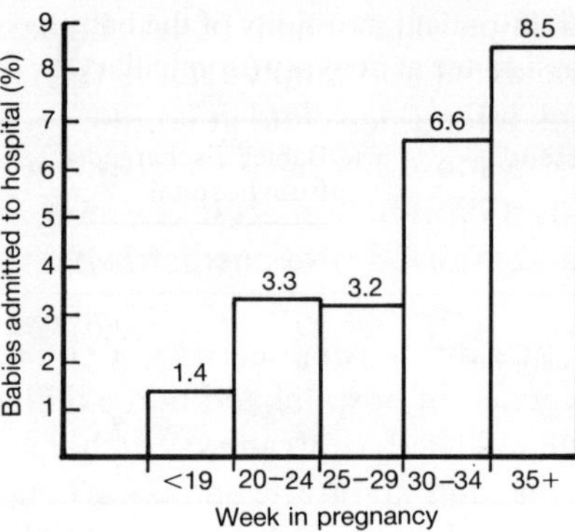

Fig. 8.4 Relationship of the week in pregnancy of midwife's first antenatal visit to subsequent in-patient morbidity of the baby in 1962.

in which the decision that hospital treatment is warranted must largely depend upon other factors (e.g. feeding difficulties, chickenpox, redundant prepuce, bronchitis). Under these circumstances few, if any, clear-cut results can be expected. However, the tables given here may form a useful basis for further analysis when a second and third year's data are available.

The finding that babies delivered at home had a similar total in-patient morbidity (3.2 per cent) to those delivered in hospital (3.5 per cent) came as a surprise. To see to what extent this was due to the special instance of a small town where paediatric beds are under the control of GPs, babies born in the town and in the local authority area surrounding it were excluded. For the remainder, the morbidity figures are 2.8 per cent for hospital deliveries and 3.1 per cent for home deliveries, so that this exclusion results in the remaining babies delivered at home actually having a *higher* in-patient morbidity than those delivered in hospital. This finding is of particular interest in view of the fact that mothers are selected for delivery in hospital in terms of many factors which might be regarded a priori as unfavourable to the infant. Using as a baseline the overall proportion of hospital deliveries of 73 per cent, the degree of selection for hospital confinement in terms of maternal age, legitimacy, and birthweight (all found to be important in this analysis) was as follows: the proportion of mothers under 20 years of age delivered in hospital was 86 per cent, of illegitimate births 91 per cent, and of babies less than 5.5 lb. 85 per cent. In other words, if hospital and home confinements were standardized for these factors, the total in-patient morbidity of babies delivered at home would have been substantially higher than for babies born in hospital. Such a finding might be due to one or both of two possibilities:

(1) That babies delivered at home are less healthy than those delivered in hospital.

(2) That their utilization of in-patient beds is higher:
   (*a*) because of social and family factors (e.g. these mothers are less able to look after a sick child at home);
   (*b*) because of a more liberal referral policy by the GP (here the policy of the GP would reflect his training and experience and the distance and availability of paediatric beds).

As far as the first possibility is concerned, a comparison of the various types of diagnoses reached in hospital and home deliveries was unrevealing (Table 8.3). The in-patient morbidity rate was higher in hospital than in home deliveries for infections, and about the same for congenital malformations and for the mixed group which includes feeding difficulties and alimentary symptoms. For trauma it was slightly lower than in home deliveries, but the number of cases was very small. The morbidity due to miscellaneous causes is slightly higher in home-delivered than in hospital-delivered babies.

In Tables 8.4 to 8.8 the total in-patient morbidity of the baby has been related to certain maternal characteristics, including age at delivery, parity, social class, legitimacy, and blood group. The morbidity has been shown to be substantially higher in babies of young mothers (under 20 years of age) than in the remainder, and this holds true for home and hospital births separately, for nulliparae and multiparae, and for Social Classes I and II, III and IV, and V separately. The difference also remains when illegitimate births are excluded. The numbers are as yet too small to permit more than very crude estimates of the relative contribution of these different factors. However, it seems that low maternal age may be more important than either high parity or low social class. Precise comparisons with the work of Morris and Heady (1955), Heady, Daly, and Morris (1955), and Daly, Heady, and Morris (1955) on neonatal (first month) and postnatal (2nd to 12th month) *mortality* are not possible, as the material reported here contains a mixture of neonatal and postneonatal events. However, it is interesting that both low maternal age and high parity were found to be important in their material. As in their results, the additive nature of these two factors is seen in the babies of multiparae under 20 years of age (Fig. 8.1c) where the morbidity rate was 7.5 per cent. On the other hand, in the babies of women under 20, social class makes little difference to morbidity (Fig. 8.1d). These trends carry the suggestion that the maternal inexperience may be an important factor in relation to the morbidity of the baby and that for young mothers, a second or third baby following close after the first is particularly at risk. The low morbidity experience of the first baby of a woman over the age of 35 (Fig. 8.1c, 1.5 per cent) makes a striking contrast to that of the second or subsequent child of the teenager (7.5 per cent).

With the single exception of birthweight (Table 8.9 and Fig. 8.3), no relationship was found between factors recorded at delivery and the subsequent morbidity of the baby in the early months. In babies whose birthweight was less than 5.5 lb., the subsequent morbidity was significantly higher than average. Douglas (1953), in a prospective study of a cohort of births, was able to detect significantly higher mortality rates in this group up to the end of the fourth year of life. The suggestion from Fig. 8.3 that the average weight may not be optimal is interesting and will be studied further.

No striking relationships were demonstrated between the subsequent in-patient morbidity of the baby and type of delivery, maternal anaesthesia, length of the second stage of labour, and administration of oxygen to the baby at birth. It may be that when sufficient numbers of cases are available to study separate classes of illness in the child (e.g. congenital malformations, respiratory infections, and so on), these analyses will prove more rewarding.

Summary

1. The in-patient hospital morbidity experience in the early months of life of 5901 babies born in 1962 to mothers resident in the Oxford Record Linkage Study area has been examined in relation to certain maternal and foetal characteristics and events at delivery. As the period of study was the calendar year 1962, the average period of exposure of the babies to the risk of admission to hospital was approximately 6 months.

2. 203 (3.4 per cent) of the babies were discharged once or more during 1962, the largest single cause of admission being congenital malformation.

3. The hospital morbidity of babies delivered at home (3.2 per cent) was almost identical to that of those delivered in hospital (3.5 per cent) in spite of the selection of difficult cases, nulliparae, and grand multiparae for delivery in hospital.

4. Of the maternal characteristics studies (age, parity, social class, and past history of stillbirth and miscarriage), low maternal age had the most important influence on the subsequent hospital morbidity of the baby.

5. Babies weighing less than 5.5 lb. at birth have a significantly higher hospital morbidity than the remainder. The lowest morbidity rate was seen in babies weighing 8 to 9 lb.

6. No definite correlations were found between type of delivery, length of the second stage of labour, type of maternal anaesthesia, or the administration of oxygen to the baby at birth, and the subsequent hospital morbidity (all causes) of the baby.

7. For domiciliary deliveries there was a striking correlation between the week of gestation in which the first antenatal visit of the midwife took place and the subsequent morbidity of the baby.

References

Acheson, E. D. (1964). Oxford Record Linkage Study. A central file of morbidity and mortality records for a pilot population. *Brit. J. prev. soc. Med.* **18**, 8.

—— and EVANS, J. G. (1964). The Oxford Record Linkage Study: a review of the method with some preliminary results. *Proc. roy. Soc. Med.* **57**, 269.

DALY, C., HEADY, J. A., and MORRIS, J. N. (1955). Social and biological factors in infant mortality, 3. The effect of mother's age and parity on social class differences in infant mortality. *Lancet* **1**, 455.

DOUGLAS, J. W. B. and MOGFORD, C. (1953). Health of premature children from birth to 4 years. *Brit. med. J.* **1**, 748.

HEADY, J. A., DALY, C., and MORRIS, J. N. (1955). Social and biological factors in infant mortality, 2. Variation of mortality with mother's age and parity. *Lancet* **1**, 395.

—— STEVENS, C. F., DALY, C., and MORRIS, J. N. (1955). Social and biological factors in infant mortality, 4. The independent effects of social class, region, the mother's age and her parity. *Lancet* **1**, 499.

KARN, M. N. and PENROSE, L. S. (1951). Birth weight and gestation time in relation to maternal age, parity and infant survival. *Ann. Eugen. (Lond.)* **16**, 147.

MORRIS, J. N. and HEADY, J. A. (1955). Social and biological factors in infant mortality, 1. Objects and methods. *Lancet* **1**, 343.

NEWCOMBE, H. B. (1964). Screening for effects of maternal age and birth order in a register of handicapped children. *Ann. hum. Genet. (Lond.)* **27**, 367.

——, KENNEDY, J. M., AXFORD, S. J., and JAMES, A. P. (1959). Automatic linkage of Vital Records. *Science* **130**, 954.

WHO (1957). 'Manual of the International Statistical Classification of Diseases, Injuries, and Causes of Death', 7th revision. Geneva.

9

The social and obstetric correlates of psychiatric admission in the puerperium

R. E. KENDELL, D. RENNIE, J. A. CLARKE, AND C. DEAN

Synopsis

Computer linkage of obstetric and psychiatric record systems made it possible to identify all women resident in the city of Edinburgh who had given birth to live or still-born children in 1971–7 and to study (a) the distribution of psychiatric admissions relative to the time of childbirth and (b) the correlates of psychiatric admission in the first 90 days after childbirth. Having a first baby, being unmarried and undergoing Caesarean section were all associated with an increased risk of admission; twin births, perinatal death and maternal age were not. It is difficult to account for these and other established relationships purely in psychological or purely in somatic terms, suggesting that influences of both kinds are probably involved in the genesis of puerperal disorders.

Introduction

It is well established that women are at considerably increased risk of admission to a psychiatric hospital in the first 3 months after childbirth (Tetlow 1955; Pugh *et al*. 1963; Paffenbarger 1964; Kendell *et al*. 1976). As most of the women concerned become psychotic within a few weeks of delivery after being symptom-free beforehand, it is almost certain that this sudden increase in hospital admissions represents a genuine rise in the incidence of psychotic illness. The cause of this increased incidence is unknown, but its magnitude implies that a potent aetiological factor is at work. Childbirth is an important life event for most women and in some circumstances can pose a major psychological stress. It is also accompanied by profound hormonal and other metabolic changes. The causal agent might therefore be either psychological or metabolic, or involve an interaction between the two. There has been much speculation on this point but as yet there is insufficient firm evidence to justify any conclusions. In principle, however, it ought to be possible to identify factors associated with an increased or decreased risk of developing a puerperal psychosis, and thereby

Reprinted from *Psychological Medicine* **11**, 341–50 (1981).

obtain important clues to the aetiology of these illnesses. The main purpose of the present investigation was to do this.

Puerperal psychoses have been studied many times since attention was first drawn to them by Esquirol (1845) and Marcé (1858). Most of these studies, however, have simply been descriptions of the clinical characteristics of the puerperal patients admitted to a particular hospital, and this approach does not allow accurate comparisons between women who did and did not become psychotic after exposure to the 'agent' childbirth. The studies of Pugh *et al.* (1963) and Paffenbarger (1964) were an important advance because they were based on all admissions from a geographically defined population; the latter also involved a group of women who did not become psychiatrically ill, but both relied on the accuracy of psychiatric records to determine whether or not the patient's illness had followed childbirth. The study of Kendell *et al.* (1976) in Camberwell differed in taking an unselected series of births rather than a series of psychotic women as its starting-point, the mothers coming to psychiatric attention being identified secondarily by searching the files of a psychiatric case register. However, as this search had to be performed by hand, the study had to be restricted to a comparatively small cohort of 2257 women. This Camberwell study can be regarded as a pilot study to the present investigation, which is based on a much larger cohort of over 35 000 women, from a defined geographical area, who delivered live or stillborn children in a 7-year period. Those who were subsequently admitted to a psychiatric hospital were identified by linking the files of the obstetric and psychiatric registers covering the area. By this means it was possible to make accurate comparisons between those who were and were not admitted to a psychiatric hospital on a wide range of obstetric and social variables.

Methods and results

Since 1 January 1970 the Royal Edinburgh Hospital has operated a case register which records psychiatric and social data in a standard format for all admissions. Since 1 January 1971 all obstetric units in Scotland have recorded extensive social and obstetric data on all hospital deliveries and returned these data on standard maternity discharge records to the Common Services Agency of the Scottish Health Service. As virtually all Edinburgh mothers have their babies in hospital, in 1 of 5 obstetric units in or near the city, and as the whole city comes within the catchment area of the Royal Edinburgh Hospital, linkage of these 2 record systems makes it possible to identify women resident within the city who have had a psychiatric admission either before or after childbirth and to compare them with those who did not on any of the variables recorded on the Maternity Discharge Record.

Between 1 January 1971 and 31 December 1977, 71 240 women delivered live or stillborn children in 1 of the 5 obstetric units in the Lothian

Region (Edinburgh and its hinterland). (Strictly speaking, these were episodes of childbirth, not individuals, as any woman having 2 children on separate occasions within the 7-year period was counted twice.) The files of the Edinburgh Psychiatric Case Register were searched for these 71 240 women. The record linkage system used is described by Heasman and Clarke (1979). It uses surname, and maiden name if applicable, initials, sex, and date of birth as identifying characteristics, and the Russell Soundex Code, which makes allowance for common misspellings, for matching surnames. For the 20 most doubtful matches, the decision to accept or reject was only made after inspection of the original psychiatric and obstetric case notes. By this means 704 women who had delivered a child in the Lothian Region in 1971–7 and also had an admission to the Royal Edinburgh Hospital in 1970–7 were identified. (Patients were identified by name only in this initial linkage process and not in subsequent analyses.)

Temporal relationship between childbirth and psychiatric admission

As information about psychiatric admissions was only available for the 8-year period from 1 January 1970 to 31 December 1977, and we wished to examine the distribution of psychiatric admissions in the 2 years before and the 2 years after childbirth, it was necessary to restrict this part of the analysis to women who had delivered a child between 1 January 1972 and 31 December 1975. The distribution of the psychiatric admissions of these mothers was calculated for 8 consecutive 90-day periods or trimesters both before and after parturition, counting each episode of childbirth separately. The results are shown in Table 9.1. As other authors have found previously, there is a dramatic rise in admissions in the first trimester after childbirth, particularly for functional psychoses. Total admissions are nearly 5 times as high as the average over the previous 2 years (43 v. 9.25), and admissions with a diagnosis of functional psychosis are no less than 16 times as high (28 v. 1.75).

Although the main post-partum peak is in the first trimester after delivery, there is a tendency for the subsequent admission rate to be higher than it was before childbirth. If the average admission rate during the 2 years before childbirth is compared with the average admission rate during the following 2 years apart from the first trimester (i.e. time periods -8 to -1 v. $+2$ to $+8$), the latter is only significantly higher, either for all admissions or functional psychoses, at the 10 per cent level. However, a CuSum plot (see Smith 1974) illustrates the change from the average admission rate in the 8 trimesters before childbirth. As Fig. 9.1 shows, the plot continues to rise steadily after the main rise in the first trimester, despite the nullifying effects of migration across the register boundary referred to below.

Table 9.1 The temporal relationship between psychiatric admission and childbirth

Time period (successive 90-day periods)	No. of psychiatric admissions	
	All diagnoses	Functional psychoses
−8	16	2
−7	9	0
−6	13	0
−5	3	1
−4	9	3
−3	7	2
−2	9	3
−1	8	3
+1	43	28
+2	10	3
+3	14	6
+4	15	3
+5	13	4
+6	15	2
+7	10	2
+8	11	1

Functional psychoses are defined as categories 295, 296, 297, 298, and 294.4 according to ICD (8th revision) and are hospital discharge diagnoses.

Relative risk of childbirth

The ratio of the number of admissions with a diagnosis of functional psychosis in the first trimester after childbirth to the average over the previous 8 trimesters $(28/1.75 = 16)$ is a measure of the 'relative risk' of childbirth, and its magnitude is of some significance. Most of the psychological stresses which have been shown to play an important role in the genesis of psychiatric disorders appear to carry a considerably lower relative risk of beween 2 and 7 (Paykel 1978). This figure of 16 is, in fact, somewhat inflated because not all the women living in the catchment area at the time of delivery would have done so throughout the preceding 2 years and any psychiatric admissions they had had while living elsewhere would have been lost, thus spuriously lowering the base rate of 1.75 admissions/trimester. But, even if 20 per cent of the women in the study had only entered the register catchment area within the previous 2 years, the corrected average admission rate over that period would rise only to 1.93, and the relative risk fall only to 14.5. And because the majority of admissions in the first trimester after childbirth take place early in this period, the relative risk would be even higher if it were calculated for a 30- rather than a 90-day risk period. The corresponding figure from the Camberwell pilot study for the relative risk of admission to

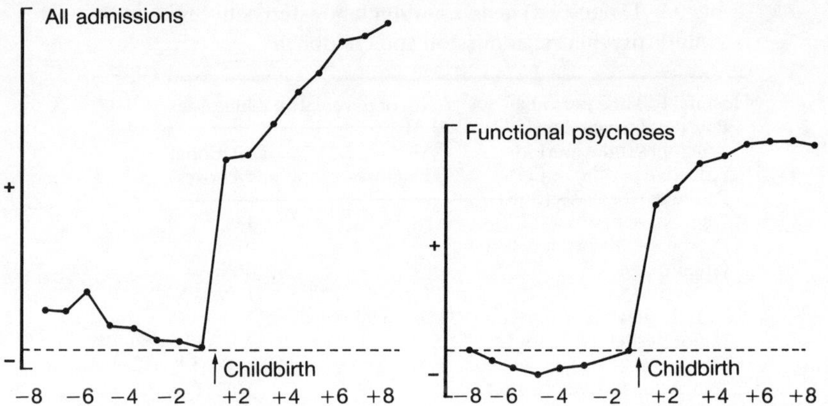

Fig. 9.1 CuSum plots of admission rates in the eight trimesters before and after childbirth.

hospital with a functional psychosis within 90 days of childbirth, adjusted for the effects of migration, was 9.7. It seems, therefore, that childbirth is a more potent influence than most other life events.

Psychiatric admissions within 90 days of childbirth

Within the group of 704 women who had both delivered a child in the Lothian Region in 1971–7 and had an admission to the Royal Edinburgh Hospital in 1970–7 were 73 who had (*a*) been living in the city of Edinburgh at the time of delivery and (*b*) been admitted to the Royal Edinburgh Hospital within 90 days of that delivery. As the analysis described above confirms that the great majority of women whose illnesses are attributable to childbirth are admitted to hospital within this period, further analyses were restricted to this group. However, as it transpired that one was a mismatch (despite having the same name, initial, and date of birth the psychiatric patient and the mother were different people) and that another had been admitted to hospital for purely social reasons (to accompany her husband, who was the real psychiatric patient), the final number was 71.

Forty of these 71 women were admitted to hospital within the first 30 days after childbirth, 18 within the next 30 days, and 13 within the last 30. So over half were admitted within the first month. Their diagnoses are given in Table 9.2, both the original hospital (discharge) diagnoses and diagnoses made according to the Research Diagnostic Criteria (RDC) of Spitzer *et al*. (1978), on the basis of information recorded in the medical and nursing case notes. The hospital diagnoses are of little interest, largely because 30 of the 71 patients were simply given a diagnosis of puerperal psychosis (in ignorance, or defiance, of the request in the British glossary to ICD-8 to avoid using the term). The RDC diagnoses demonstrate the predominantly

Table 9.2 Diagnoses of the 71 women admitted within 90 days of childbirth

Hospital (discharge) diagnoses	No.	
Puerperal psychosis (ICD-8 294.4)	30	
Schizophrenia (295)	5	
Affective psychosis (296)	8	
Reactive psychosis (298)	2	
Depressive neurosis (300.4)	11	
Personality disorder (301)	9	
Other diagnoses	6	
Research Diagnostic Criteria (RDC) diagnoses		
Major depressive disorder	33	(19 definite, 14 probable)
Minor depressive disorder	16	(13 definite, 3 probable)
Manic disorder	8	(all definite)
Hypomanic disorder	1	(probable)
Schizo-affective disorder	4	(all definite)
Schizophrenia	1	(probable)
Unspecified functional psychosis	5	
Other diagnoses	3	

affective nature of these illnesses, 49 having major or minor depressive disorders and 9 manic or hypomanic disorders. Not one patient met the RDC criteria for schizophrenia, though 4 patients fulfilled the criteria for definite schizo-affective disorder and 1 woman with a major depressive disorder had a probable schizophrenic defect state as well. The patient with probable schizophrenia had had previous episodes unrelated to childbirth. There was no significant difference in the distribution of either hospital or RDC diagnoses between patients admitted within the first 30 days after childbirth and those admitted between 30 and 90 days, though the manic patients tended to be admitted early.

Correlates of psychiatric admission within 90 days of childbirth

The 71 women described above were derived from a known population—the 35 729 female residents of the city of Edinburgh who delivered babies in 1971–7—and can safely be assumed to constitute all, or virtually all, the members of that population who were subsequently admitted to a psychiatric hospital, though not, of course, all those who became psychiatrically ill. (Again, it is necessary to remember that these are 35 729 episodes of childbirth, not individuals, though it so happened that all the 71 episodes followed by psychiatric admission involved different individuals.) The ready availability of the maternity discharge records of these 35 729 women makes it possible to determine which of a wide range of obstetric and social variables is associated with an increased, or decreased, risk of psychiatric admission.

The incidence of all the potentially relevant variables in these records was therefore determined both in the 71 patients and the 35 729 controls and the 2 groups were compared. Parallel comparisons were also carried out for patients with psychotic illnesses (arbitrarily defined as all patients fulfilling RDC criteria for major depressive, manic, or schizo-affective disorder, schizophrenia, or unspecified functional psychosis) and with depressive illness (RDC major and minor depressive disorder). The results are summarized in Tables 9.3 and 9.4.

The 71 psychiatric patients differed significantly from the controls on surprisingly few of these variables. More of them were single, or widowed, divorced, or separated at the time of delivery, and correspondingly fewer were married. More were having their first baby, and more were delivered by Caesarean section. However, as Table 9.4 shows, they did not differ significantly from the controls in the incidence of any of the numerous physical complications of pregnancy or delivery that were examined, or in most aspects of their previous obstetric histories. Nor was there any difference between patients and controls in maternal age, length of gestation, or the sex or birthweight of the baby. The pattern is similar both for the 51 patients with psychotic illnesses and for the 49 with depressive illnesses, though it may be important (see Table 9.3 and the Discussion below) that the relationship with marital status is weaker, and in the case of the psychotic patients no longer statistically significant.

Although the maternity discharge record contains comprehensive obstetric information, it contains relatively little background information about the mother. The psychiatric case register contains more information of this kind and, although it was impossible to compare the psychiatric patients with the 35 729 obstetric controls on ratings obtained from this source, it was possible to compare them with other psychiatric patients. As the majority of the 71 puerperal patients had hospital diagnoses of some form of functional

Table 9.3 Social and obstetric variables derived from the maternity discharge records associated with an increased psychiatric risk

Variable	Psychiatric patients ($N = 71$) No. (%)	Psychotic patients ($N = 51$) No. (%)	Depressed patients ($N = 49$) No. (%)	Control population ($N = 35\ 729$) %
Marital state of mother				
Single	10 (14.1)	5 (9.8)	6 (12.2)	7.3
Married	56 (78.9)**	42 (82.3)	39 (79.6)*	90.9
Widowed, divorced, separated	5 (7.0)	4 (7.8)	4 (8.2)	1.8
No previous pregnancies	36 (50.7)	33 (64.7)**	30 (61.2)**	39.4
No living children	44 (62.0)*	33 (64.7)*	30 (61.2)	46.8
Caesarean section	12 (16.9)*	11 (21.6)**	9 (18.4)*	8.4

  * Indicates probability $p < 0.05$.
** Indicates probability $p < 0.01$.

Table 9.4 Social and obstetric variables not significantly associated with psychiatric admission

Variable	Psychiatric patients ($N = 71$) No. (%)	Control population ($N = 35\ 729$) %	Variable	Psychiatric patients ($N = 71$) No. (%)	Control population ($N = 35\ 729$) %
Age of mother at delivery			Birthweight of baby (g)		
Under 20 years	6 (8.5)	11.6	Under 2000	1 (1.4)	2.2
20–29	50 (70.4)	68.3	2000–2999	15 (21.1)	22.1
30–39	15 (21.1)	18.9	3000–3999	51 (71.8)	67.6
40 or over	0	1.2	4000 or over	4 (5.6)	8.1
Past obstetric history			Outcome of pregnancy		
* Rhesus iso-immunization	0	0.3	Multiple births	0	1.0
* Eclampsia/hypertension	1 (2.9)	6.1	Perinatal death	2 (2.8)	1.7
* Ante-partum haemorrhage	0	2.0	Forceps delivery	12 (16.9)	18.7
* Multiple births	1 (2.9)	0.5	** Sterilization afterwards	1 (2.8)	2.2
* Post-partum haemorrhage	1 (2.9)	3.3			
* Stillbirth	0	1.8	Complications of pregnancy		
* First year death	0	4.3	Ante-partum haemorrhage	3 (4.2)	2.3
** Spontaneous abortion	5 (13.9)	16.1	Anaaemia of pregnancy	1 (1.4)	3.2
** Therapeutic abortion	3 (8.3)	6.0	Malposition *in utero*	2 (2.8)	0.9
Caesarian section	4 (5.6)	3.3	Hydramnios	0	1.1
** Perinatal death	3 (8.3)	3.5	Renal/urinary infection	6 (8.4)	6.4
			Pre-eclampsia	5 (7.0)	11.3
** Gestation period					
Under 38 weeks	3 (8.4)	10.3	Complications of delivery		
38–39 weeks	14 (38.9)	31.6	Post-partum haemorrhage	0	4.9
40–41 weeks	18 (50.0)	53.6	Abnormality of bony pelvis	0	1.2
42 or over	1 (2.8)	4.6	Malpresentation of foetus	2 (2.8)	1.3
			Prolonged labour	1 (1.4)	3.5
Sex of baby			Laceration of perineum	2 (2.8)	3.5
Male	37 (52.1)	51.3	Sepsis	2 (2.8)	3.4
Female	34 (47.9)	48.7	Pyrexia of unknown origin	1 (1.4)	1.8
			Puerperal anaemia	2 (2.8)	1.8

The format of the maternity discharge record was changed in January 1975. Items marked * were only recorded in 1971–4 and those marked ** only recorded in 1975–7. Thirty-five psychiatric patients were admitted in 1971–4 and 36 in 1975–7.

psychosis, the 802 women aged 15–39 who had been admitted to the Royal Edinburgh Hospital with functional psychoses between 1 January 1971 and 31 December 1977 were used as a second control population. Three variables were studied, and the results are shown in Table 9.5. Place of birth was examined as an indirect indication of the likelihood of the mother being isolated from her family and culture of origin. In the event, however, neither the puerperal patients as a whole nor their psychotic or depressed subgroups differed significantly from other psychiatric patients in either place of birth or social class. They did differ markedly in whom they were living with at the time of admission, a far higher proportion living with a spouse or cohabitee (78 per cent v. 33 per cent and $p < 0.001$) and a correspondingly lower proportion with their parents or elsewhere, but this is probably just a reflection of the fact that pregnant women are more likely to be living with a husband or other sexual partner than women of childbearing age in general.

Discussion

The main purpose of this study was to identify obstetric and social variables associated with an increased, or decreased, risk of puerperal psychosis.

Table 9.5 Differences between puerperal patients and 802 women aged 15–39 admitted to the same hospital during the same time period with non-puerperal psychoses

Variable	Puerperal patients ($N = 71$) No. (%)	Psychiatric controls ($N = 802$) (%)
Place of birth		
Edinburgh	35 (49.3)	56.7
Rest of Scotland	14 (19.7)	19.1
Rest of UK	13 (18.3)	12.2
Outside of UK	6 (8.5)	9.5
Not known	3 (4.2)	2.5
Social class		
I or II	19 (26.8)	24.2
III	24 (33.8)	35.9
IV or V	19 (26.8)	25.7
Other	9 (12.6)	14.2
Currently living with		
Spouse or cohabitee	55 (77.5)	32.9
Parent(s)	6 (8.5)	37.3
Other relatives or friends	5 (7.0)	14.6
Lodgings or institution	3 (4.2)	6.2
Alone	2 (2.8)	7.1

Three such variables have been identified—parity, marital status, and mode of delivery.

'No previous pregnancies' is associated with a greatly increased risk of both psychotic and depressive illness, and 'no living children' with an increased risk of psychiatric admission and of psychotic illness. It is not possible to decide which is more important because the two are so closely related. (For all the women who developed psychotic ($N = 33$) or depressive ($N = 30$) illnesses after their first confinement, this was also their first pregnancy.) Overall, there were 71 puerperal admissions from 35 729 confinements, so the risk is 2.0/1000 births, a figure that has been obtained several times before (Hemphill 1952; Paffenbarger 1964; Grundy and Roberts 1975). The 16 716 first confinements resulted in 44 admissions, a risk of 2.63/1000, whereas the 19 013 later confinements resulted in only 27 admissions, a risk of 1.42/1000. The risk of psychiatric admission in multiparae is therefore only 54 per cent of that in primiparae.

It is well known that women who have had one puerperal psychosis have a high risk of becoming psychotic again if they have further children. Vislie (1956), Foundeur *et al*. (1957), and Protheroe (1969) all found the risk after further pregnancies to be 1 in 7 and Paffenbarger (1964) found it to be 1 in 3. As many women must be deterred from having further children by this high risk of recurrence, it is important to consider whether the observed difference in the risk of psychiatric admission after first and subsequent children could be due simply to the avoidance of further children by women who had had the misfortune to develop a psychosis after their first.

Of the 35 729 births to Edinburgh mothers in 1971–7, 16 716 were first babies and 19 013 non-first. Assuming that the age at which women have their children is not changing, it follows that the average number of further episodes of childbirth experienced by Edinburgh women having at least one child is

$$\frac{35\,729 - 16\,716}{16\,716} = 1.14$$

Let us assume that every one of the 44 women who had a psychiatric admission after her first child avoided all further pregnancies, that their risk of a further puerperal psychosis would have been 1 in 7, and that for the population as a whole the real risk of a psychiatric admission after subsequent children is the same as after the first. Under these circumstances the 19 013 non-first deliveries would generate

$$\frac{19\,013 \times 2.63}{1000} - \frac{44 \times 1.14}{7} = 42.83 \text{ psychiatric admissions,}$$

and the observed risk of psychiatric admission after non-first deliveries would be

$$\frac{42.83 \times 1000}{19\,013} = 2.25/1000$$

This is much higher than the observed risk of 1.42/1000. Even if we assume, following Paffenbarger, that the risk of a further puerperal psychosis is 1 in 3 rather than 1 in 7, the 19 013 non-first deliveries would still generate

$$\frac{19\,013 \times 2.63}{1000} - \frac{44 \times 1.14}{3} = 33.28 \text{ psychiatric admissions}$$

and an observed risk of psychiatric admission of 1.75/1000. In other words, even on the extreme assumption that no woman who has a psychiatric admission after her first child ever risks having any further children, and that the risk of a second puerperal psychosis is as high as 1 in 3, the difference between the observed admission rates after first and subsequent children cannot be explained without assuming that first episodes of childbirth are inherently more likely to be followed by psychiatric admissions than subsequent episodes.

In fact, the observation that puerperal psychoses are commoner in primiparae than in multiparae was made repeatedly in days when it was far harder for women to avoid unwanted pregnancies than it is now. Thomas and Gordon (1959) combined the findings of 13 American and British studies published between 1913 and 1958, a total of 1100 cases, and showed that 54 per cent were primiparae. After allowing for average family size in those decades they concluded that 'there appears little doubt that the condition is primarily one of primiparae'.

The second variable associated with an increased risk of psychiatric admission is marital status. For all 71 admissions there was a statistically significant relationship both with being single ($P < 0.05$) and, independently, with being widowed, divorced, or separated ($P < 0.01$). The relationship with marital status was less strong, however, for the 51 psychotic patients ($P < 0.1$). Twenty-nine of the 95 women on whom Esquirol's original description of puerperal psychosis was based were unmarried, and an increased risk of puerperal psychosis in unmarried women has been observed by others more recently (for example, Tetlow 1955), but this has not been a consistent finding. The relationship with marital status observed here is not secondary to a relationship with maternal age because no such relationship was found. Women without a husband to support them are, however, probably more likely to be admitted to hospital than other young mothers. It cannot be assumed, therefore, that the observed relationship with marital status is necessarily a reflection of genuine differences in incidence, particularly as it fell short of statistical significance in the 51 patients with the most severe illnesses. Much depends on whether or not a similar excess of psychopathology in unmarried mothers is present in women who are not admitted to hospital, and we hope to have evidence on this point shortly.

The third important variable is mode of delivery. Although forceps delivery was not associated with any increased risk, delivery by Caesarean section was so associated, particularly in the 51 psychotic patients ($P < 0.01$). Before 1975, elective and emergency operations were coded together, but seven of the eight psychiatric admissions associated with Caesarean section in 1975–7 were after elective operations.

Having a first baby, being delivered by Caesarean section, and having a baby when unmarried are all situations in which one would expect the psychological stress of childbirth to be particularly severe. We also have evidence that this commonsense view is shared by puerperal women themselves. We gave a questionnaire to 85 unselected women in a post-natal ward a few days after delivery, asking them to rate on 10 cm linear scales 'how they imagined most mothers would feel' in the six situations listed in Table 9.6, As the mean scores shown in that table indicate, these women, who had recently experienced the stresses of childbirth themselves, expected a first baby to be more stressful than a second or third, and a Caesarean section or having a baby when unmarried to be more stressful still. However, these findings do not allow us to conclude that the aetiological role of childbirth is due entirely, or even mainly, to its significance as a 'life event' and that hormonal and other metabolic changes are unimportant, for there are a number of facts which do not fit this simple model.

The most obvious is that women who have had one puerperal psychosis are at high risk—at least 1 in 7—of having further psychoses after subsequent children, which is cogent evidence for some kind of constitutional predisposition. A second problem is the absence of any increased risk of puerperal psychosis after twin births, stillbirths, or neonatal deaths, or after a history of such events in earlier pregnancies, despite the commonsense view,

Table 9.6 Estimates of 85 post-partum women of the stress involved in six different obstetric experiences

Obstetric event	Mean rating on a 10 cm linear scale
(a) Having a second or third baby	29.9
(b) Having a first baby	43.4
(c) Having a baby delivered by Caesarean section	57.1
(d) Having a baby without being married	61.4
(e) Having twins	64.7
(f) Having a baby which was born dead or died soon after birth	88.4

All differences are statistically significant at the 5% level ($p < 0.05$) apart from those between c and d, and between d and e.

shared by the 85 mothers we consulted, that these events should be particularly stressful psychologically. The Camberwell pilot study also failed to reveal any increased risk after stillbirths or twins. It could be argued, though, that there is a threshold beyond which increased stress does not produce any further increase in morbidity, and that twin births and perinatal deaths come beyond that threshold, or that stillbirths and twin births are simply too rare for statistically significant differences to emerge without much larger numbers. (Although there were 623 perinatal deaths and 359 twin births in this series, they might well have failed to evoke more than two or three psychoses even if the true risk were double the risk of 2/1000 after live-births.) The literature on this point is inconclusive. Hemphill (1952) failed to find any increased risk of psychiatric admission after twin deliveries, but Paffenbarger (1964) did find an increased risk after perinatal deaths.

The evidence suggesting that there is no increased psychiatric morbidity in fathers after childbirth (Kendell *et al*. 1976), or in women after therapeutic abortion (Tietze and Lewit 1972; Brewer 1977), is a further problem, for the birth of a son or daughter must be an important life event for many fathers, and a termination likewise for many women. A hormonal theory of puerperal psychoses, however, can easily accommodate both findings, for hormonal changes are modest after termination of early pregnancy, and non-existent in fathers. Moreover, most puerperal illnesses are depressive and there is good evidence that the psychological events precipitating depression are predominantly losses, or exists (Paykel *et al*. 1969), yet childbirth cannot be construed as a loss in any straightforward way. Finally, the relative risk of psychotic illness after childbirth seems to be considerably higher than that attributable to other life events, suggesting that a mechanism of a different type is involved.

On the other hand, the evidence that the hormonal and other metabolic changes taking place during and after childbirth in women who develop either mild mood disturbances or fully fledged psychoses are any different from those of normal women is weak and inconclusive (Nott *et al*. 1976; Ballinger *et al*. 1979; Handley *et al*. 1980). Neither were we able to confirm the relationships between puerperal psychosis and maternal age, gestation period, and birthweight reported by Paffenbarger (1964) which led him to conclude that these illnesses were 'probably somatic in origin and related to endocrine imbalance'. In summary, therefore, the relationships observed here with parity, marital status, and Caesarean section are all readily explicable in psychological terms. The increased risk after Caesarean section could easily be explained in metabolic terms instead, and the relationship with marital status might simply be a nosocomial effect, but it is harder to account for the increased incidence in primiparae without invoking psychological factors. This does not mean, however, that the dramatic hormonal and other physiological changes which occur in the aftermath of parturition play no part in the genesis of puerperal psychoses. The suspicion that they do so must remain even if firm evidence to that effect is lacking.

In the past there has been much discussion and disagreement about the nature of the psychoses developing in the puerperium. At one time it was widely believed that these illnesses had a distinct psychopathology of their own, and a distinct aetiology and prognosis as well. Although interest in this possibility still persists (see, for example, Brockington *et al.* 1978) the predominant contemporary view is that most puerperal illnesses are simply schizophrenic or affective psychoses precipitated by childbirth, and that their treatment and prognosis are not materially different from similar illnesses occurring under other circumstances. The relative preponderance of schizophrenic and affective syndromes varies greatly in different published series. Thomas and Gordon (1959) compared 12 American and British studies published between 1911 and 1958, and the proportion of schizophrenic illnesses varied from 14 to 65 per cent and of manic depressive illnesses from 15 to 70 per cent. In view of what we now know of the instability of diagnostic criteria from one decade to the next, and of gross Anglo-American differences in the 1940s and 1950s, little significance can be attached to any of these percentages. To our knowledge the 71 patients in this study are the first representative series of puerperal admissions to be diagnosed using operational criteria and the absence of schizophrenic illnesses is very striking. As Table 9.2 shows, over 80 per cent fulfilled RDC criteria for either depressive or manic disorders. Not a single patient fulfilled criteria for definite schizophrenia and the only one to qualify as probable schizophrenia had had previous episodes unrelated to childbirth. The results of the Camberwell pilot study were the same. The only schizophrenic illness encountered there was in a woman who had had two previous episodes unrelated to childbirth. It is well known that in women with a history of manic depressive illnesses childbirth is likely to precipitate further episodes. Bratfos and Haug (1966) estimated the risk as 1 in 5, Reich and Winokur (1970) as nearly 1 in 3. It seems likely, therefore, that the great majority of puerperal psychoses are basically affective disorders, though carefully matched comparisons with non-puerperal controls and follow-up studies will be needed to confirm this.

The CuSum plots shown in Fig. 9.1 suggest that the incidence, or at least the admission rate, of psychiatric disorders remains higher than it was before childbirth for at least 2 years after the event, and that the post-partum rise is not restricted to the first 90 days. Although the before/after difference is only statistically significant at the 10 per cent level, it is unlikely to be a chance finding because exactly the same was found in the Camberwell pilot study. Puerperal illnesses, both those which are severe enough to require hospital admission and those which are not, are predominantly depressive in nature and it is known that the mothers of young children have a high incidence of depressive symptoms (Brown and Harris 1978; Richman 1978). It is widely appreciated that the nature of the relationship between puerperal psychoses and the much commoner 'post-partum blues' is an important unresolved issue. The finding of a raised psychiatric admission rate for at least 2 years after childbirth suggests that the relationship between puerperal depressions and the depressions of young mothers may be an equally important subject for future study.

Acknowledgements

We are grateful to the staff of the Scottish Office Computer Service for carrying out the linkage between the maternity discharge records and psychiatric records, and to Mr D. Denoon for his technical assistance.

References

BALLINGER, C. B., BUCKLEY, D. E., NAYLOR, G. J., AND STANSFIELD, D. A. (1979). Emotional disturbance following childbirth: clinical findings and urinary excretion of cyclic AMP. *Psychological Medicine* **9**, 293–300.

BRATFOS, O. AND HAUG, J. O. (1966). Puerperal mental disorders in manic depressive females. *Acta psychiatrica scandinavica* **42**, 285–94.

BREWER, C. (1977). Incidence of post-abortion psychosis: a prospective study. *British Medical Journal* **i**, 476–7.

BROCKINGTON, I. F., SCHOFIELD, E. M., DONNELLY, P., AND HYDE, C. (1978). A clinical study of post-partum psychosis. In *Mental Illness in Pregnancy and the Puerperium* (ed. M. Sandler) pp. 59–68. Oxford University Press, Oxford.

BROWN, G. W. AND HARRIS, T. (1978). *Social Origins of Depression*. Tavistock Publications, London.

ESQUIROL, E. (1845). *Mental Maladies: a Treatise on Insanity* (trans. E. K. Hunt). Lea and Blanchard, Philadelphia.

FOUNDEUR, M., FIXSEN, C., TRIEBEL, W. A., AND WHITE, M. A. (1957). Postpartum mental illness: a controlled study. *Archives of Neurology and Psychiatry* **77**, 503–12.

GRUNDY, P. F. AND ROBERTS, C. J. (1975). Observations on the epidemiology of postpartum mental illness. *Psychological Medicine* **5**, 286–90.

HANDLEY, S. L., DUNN, T. L., WALDRON, G., AND BAKER, J. M. (1980). Tryptophan, cortisol and puerperal mood. *British Journal of Psychiatry* **136**, 498–508.

HEASMAN, M. A. AND CLARKE, J. A. (1979). Medical record linkage in Scotland. *Health Bulletin* **37**, 97–103.

HEMPHILL, R. E. (1952). Incidence and nature of puerperal psychiatric illness. *British Medical Journal* **ii**, 1232–5.

KENDELL, R. E., WAINWRIGHT, S., HAILEY, A., AND SHANNON, B. (1976). The influence of childbirth on psychiatric morbidity. *Psychological Medicine* **6**, 297–302.

MARCÉ, L. V. (1858). *Traité de la Folie des Femmes Enceintes, des Nouvelles Accouchées et des Nourices*. Baillière et fils, Paris.

NOTT, P. M., FRANKLIN, M., ARMITAGE, C., AND GELDER, M. G. (1976). Hormonal changes and mood in the puerperium. *British Journal of Psychiatry* **128**, 379–83.

PAFFENBARGER, R. S. (1964). Epidemiological aspects of parapartum mental illness. *British Journal of Preventive and Social Medicine* **18**, 189–95.

Paykel, E. S. (1978). Contribution of life events to causation of psychiatric illness. *Psychological Medicine* **8**, 245–53.

——, Myers, J. K., Dienelt, M. N., Klerman, G. L., Lindenthal, J. J., and Pepper, M. P. (1969). Life events and depression: a controlled study. *Archives of General Psychiatry* **21**, 753–60.

Protheroe, C. (1969). Puerperal psychoses: a long term study 1927–1961. *British Journal of Psychiatry* **115**, 9–30.

Pugh, T. F., Jerath, B. K., Schmidt, W. M., and Reed, R. B. (1963). Rates of mental disease related to childbearing. *New England Journal of Medicine* **268**, 1224–8.

Reich, T. and Winokur, G. (1970). Postpartum psychoses in patients with manic depressive disease. *Journal of Nervous and Mental Disease* **151**, 60–8.

Richman, N. (1978). Depression in mothers of young children. *Journal of the Royal Society of Medicine* **71**, 489–93.

Smith, A. (1974). Some new techniques for epidemic monitoring. *Medicine* **35**, 2052–4.

Spitzer, R. L., Endicott, J., and Robins, E. (1978). *Research Diagnostic Criteria for a Selected Group of Functional Disorders* (3rd edn). New York State Psychiatric Institute, New York.

Tetlow, C. (1955). Psychoses of childbearing. *Journal of Mental Science* **101**, 629–39.

Thomas, C. L. and Gordon, J. E. (1959). Psychosis after childbirth: ecological aspects of a single impact stress. *American Journal of the Medical Sciences* **238**, 363–88.

Tietze, C. and Lewit, S. (1972). Joint program for the study of abortion: early medical complications of legal abortion. *Studies in Family Planning* **3**, 97–119.

Vislie, H. (1956). Puerperal mental disorders. *Acta psychiatrica et neurologica scandinavica* Suppl. 111.

10

Record linkage for drug monitoring
D. C. G. SKEGG AND R. DOLL

Summary

A study was carried out to assess the feasibility of using record linkage for drug monitoring. For two years, three types of records were collected for a total of 43,117 people: (1) details of basic attributes, such as sex and age; (2) details of prescriptions dispensed; and (3) records of hospital admissions, obstetric deliveries, and deaths. The records about each person were linked together, and analyses were performed to reveal associations between drugs and diagnoses. The study suggested that record linkage would be useful both for generating and for testing hypotheses about the adverse effects of drugs. The method would be especially valuable for detection of delayed effects (such as the induction of cancer), sudden deaths outside hospital, and effects on the fetus—all of which are difficult to study by other means. A full-scale project would need to cover a larger population, and some of the practical issues that would arise are discussed.

All drugs can produce unwanted effects and, with the great advances in therapeutics during the past few decades, iatrogenic disease has become a major challenge to preventive medicine. Because adverse drug reactions are generally not unique clinical entities, the monitoring of drug-induced disease is not so much a matter of clinical diagnosis but more an epidemiological task of detecting associations between drugs and events (where an 'event' is any untoward happening experienced by a patient) (Finney 1965). Further research is then needed to determine which associations are due to drug toxicity and, in such cases, to define the magnitude of risk.

In the past, most drug monitoring schemes have relied on clinicians to recognize associations between drugs and events. Doctors are likely to suspect an association if the drug causes the event commonly and without delay, especially if the event happens to be one of those frequently induced by drugs, such as rashes and blood dyscrasias. Clinical suspicion is unlikely, on the other hand, if the reaction occurs rarely or after a long delay, or if it is not typical of drug-induced disease. There is an urgent need for improved methods of drug monitoring that would be capable of detecting unsuspected effects, especially delayed ones, such as the induction of cancer, autoimmune disease, and nephropathies.

Reprinted from *Journal of Epidemiology and Community Health* **35**, 1, 25–31 (1981).

One suitable approach would be to record the medicines given to members of a large population and then to study their morbidity and mortality over a long period. Prospective surveillance of this kind would be a daunting task unless it could be achieved by linking records collected routinely for other purposes. In Britain the vast majority of prescriptions are written in general practice and are collected centrally for pricing. If these prescriptions could be linked with records of morbidity and mortality, there could be an excellent opportunity to detect major adverse effects of drugs. We carried out a study to assess the feasibility of using record linkage for this purpose.

Methods

The study population consisted of the people registered with 20 general practitioners (GPs) in six practices. In one group practice, the methods assessed differed from the others and are reported elsewhere (Skegg, Richards, and Doll 1981). The remaining practices were in the Oxford region and comprised four group practices and one solo practice (included only for the second year of the study). The 16 GPs were responsible for about 33 000 people at any one time, and altogether 43 117 people were included during the two years from 1 March 1974 to 29 February 1976.

Information obtained

During this two-year period, three main types of record were obtained for every person in the population:

Basic attributes
These included name (and title), address, sex, date of birth, date of registration with the practice, and date of removal. This information was obtained from computer records maintained by the Oxford Community Health Project (Dunleavy and Perry 1977), which had already assigned to each person a unique number (the 'person number') consisting of six digits and a check character.

Prescriptions for drugs
The Prescription Pricing Authority provided photocopies of all prescriptions bearing the stamps of the GPs. This method of data collection involved no work for the doctors and ensured that only prescriptions actually dispensed were recorded. Prescriptions written by medical assistants and locums were obtained, because these doctors have to use the prescription forms of the principals. Drugs prescribed by hospital doctors, however, were not included.

The relevant person number was assigned to each prescription. Although the identifying data on prescriptions are limited, each prescription has a GP's name and address stamped on it, and identification becomes much simpler when the data can be compared with a list of the names, addresses, and ages of people in one practice. Identification was also simplified by asking doctors to add the person numbers, which were displayed on case-notes, to prescriptions when convenient, and to identify temporary residents. Using these approaches, we needed to refer very few prescriptions to administrative staff at the practices.

Each prescription was coded and the following information was keyed into the computer: the person number, the date of the prescription, the name of the drug and its form (for example, capsule or lotion), the name of the doctor responsible for the prescription, and—for some practices—the status of the person who actually wrote it.

To validate the method of obtaining prescriptions from the pricing authority, the GPs agreed to make carbon copies of all the prescriptions they wrote during sample periods, and to record the indications for drugs at the same time. During such periods, the number of prescriptions returned by the pricing authority was actually slightly higher than the number recorded in the practices (5553 against 5248), despite the fact that some prescriptions would not have been dispensed. The difference was due to the failure of GPs to make copies of all prescriptions.

Records of morbidity and mortality
The Oxford Record Linkage Study (Baldwin 1972) provided details of the following events involving patients in the participating practices: deaths (in or out of hospital); general hospital admissions (with discharge diagnoses); psychiatric in-patient, day-patient, and out-patient care; admissions to maternity hospitals (at all stages of pregnancy); and obstetric deliveries. The identifying information on each event record was used to obtain the patient's person number; since the number of events was relatively small, this was done by hand.

Record linkage

Using an ICL 1900 series computer, the records about each person's basic attributes, the drugs he or she received, and his or her morbidity or mortality were linked by means of the person number. The records about each person were stored in chronological order. Table 10.1 shows the total numbers of records in the main consolidated file. A smaller file contained data about 699 women who were pregnant during the study. These files contained no names or addresses, and particular care was taken to safeguard confidentiality.

Analysis

The main consolidated file was analysed to reveal associations between drugs and events. Medicines were automatically recoded according to the

Table 10.1   Total numbers of records in
the main consolidated file

Type of record	No.
Header (basic attributes)	43 117
Prescription	274 453
Hospital admission*	4 982
Psychiatric in-patient	190
Psychiatric day-patient	89
Psychiatric out-patient	145
Death	398

* Psychiatric and maternity admissions
not included.

approved names of their constituents, and diagnoses were classified in the three-digit categories of the International Classification of Diseases (8th revision). The people who received a particular drug were regarded as 'users'. For each drug, the frequency of each diagnosis among users was compared with the frequency among non-users (that is, the remainder of the population). The association between the drug and the diagnosis was defined by calculating a 'relative risk' (without implying a causal relationship) and a χ^2 value. The relative risk and χ^2 were adjusted for sex and age by stratifying the data and applying the method of Mantel and Haenszel (1959). The results of this screening analysis were tabulated in two computer listings, one showing the diagnoses associated with each drug, the other showing the drugs associated with each diagnosis; statistically significant associations were marked by asterisks.

The preliminary analysis did not distinguish between different types of event, for example, hospital admission or death. These could be separated in further analyses that were carried out for associations of interest. At the same time, it was possible to combine different drugs (for example, all benzodiazepines) or different diagnoses (for example, all types of cerebrovascular disease) and it was also possible to control for the use of other types of medicine.

Another factor not considered in the preliminary analysis was the time relationship between the prescribing of a drug and the recording of a diagnosis. Although this factor could have been considered in the analysis, the number of patients in this study with a particular combination of drug and diagnosis was usually so small that it was preferable to examine the case histories individually. A computer program was used to print out all the records of such patients in chronological order, showing, in plain language, the drugs they received (including the brand-names and dosage forms) and details of hospital admissions, psychiatric care, and deaths. Interesting associations were often studied further by examining patients' general practice or hospital case-notes.

The file of maternity data was analysed by similar methods. We searched for associations between drugs prescribed during pregnancy and diagnoses in the mothers, and for associations between drugs prescribed during pregnancy and diagnoses in the babies. Separate analyses were performed for drugs precribed during early pregnancy and for those prescribed at any stage. Since the number of pregnancies in this feasibility study was too small to provide useful information about adverse effects of drugs, the results of these analyses will not be presented.

Results

Analysis of the main file revealed several thousand statistically significant associations between drugs and diagnoses, the majority of which were predictable. In many instances, the drug was known to be a treatment for the disease. This explained associations between thyroxine and myxoedema ($\chi^2_1 = 156.0$), and between chlorpromazine and schizophrenia ($\chi^2_1 = 721.3$). In other cases, the drug was known to be a treatment for some complication of the disease. Such an explanation could account for associations between kaolin, a constituent of antidiarrhoeal mixtures, and cancer of the rectum ($\chi^2_1 = 16.4$), and between phenytoin, an anticonvulsant, and cerebral malignancy ($\chi^2_1 = 124.8$). Medicines were sometimes prescribed to prevent or treat the unwanted effects of other drugs: treatment for extrapyramidal effects of phenothiazines accounted for an association between benzhexol and schizophrenia ($\chi^2_1 = 875.0$).

Associations due to treatment could not be eliminated by confining the analysis to drugs prescribed before recording of a diagnosis, because the treatment of most diseases is begun before patients are admitted to hospital, and many admissions are not the first for a particular condition.

Some significant associations that were not due to treatment probably resulted from confounding. For example, an association between bromhexine (an expectorant) and cerebral thrombosis ($\chi^2_1 = 7.2$) may have been due to the fact that both chronic bronchitis and cerebral thrombosis are more common among smokers (Doll and Peto 1976); an association between metronidazole, a trichomonacide, and legal abortion ($\chi^2_1 = 5.1$) can be attributed to the fact that both trichomoniasis and abortion are associated with promiscuity; and an association between dioctyl sodium sulphosuccinate, a faecal softening agent, and chronic ulceration of skin ($\chi^2_1 = 14.6$) probably resulted from the fact that both constipation and bedsores are complications of immobility.

Of the large number of associations that were not explained by considerations such as these, many were studied further in an attempt to determine whether drug toxicity was a likely explanation. Two examples will be described here.

Digoxin and diarrhoea

Among 637 patients treated with digoxin, five were admitted to hospital with a diagnosis of diarrhoea, whereas the number expected from the experience of 42 480 other people, adjusting for sex and age, was 0.8. This difference was highly significant (Table 10.2).

Examination of case-notes showed that, in four cases, the time sequence and clinical picture were consistent with the assumption that digoxin caused the diarrhoea. Nausea and vomiting were also prominent, except in one case. Three of these patients were over 80, and the other was a girl aged three months.

The tendency for preparations of foxglove to cause diarrhoea has been well known since the 18th century (Withering 1785). Although usually associated with nausea and vomiting, diarrhoea can be the only gastrointestinal manifestation of digitalis toxicity (Moe and Farah 1975). The risk of intoxication is particularly high in the elderly and in children.

Table 10.2 Frequency of admission to hospital with a diagnosis of diarrhoeal disease (International List 009), according to use of digoxin

Use of digoxin	No. with diarrhoea	Total no. in study	Relative risk*
Users	5	637	6.1
Non-users	24	42 480	

* Adjusted for sex and age, according to the method of Mantel and Haenszel (1959). $p < 0.01$.

Aspirin and cerebral haemorrhage

There was a statistically significant association between prescribed aspirin and cerebral haemorrhage: three aspirin users were admitted to hospital with cerebral haemorrhage, compared with an expected number of 0.4 (Table 10.3). In each case, the patient died and the diagnosis was confirmed at necropsy. The frequency of other kinds of cerebrovascular disease in users of aspirin was no different from that in the remainder of the population.

This finding is a cause for concern, since aspirin is known to impair haemostasis in several ways (Prescott 1975), but it must be interpreted with caution. The observation is based on only three cases and could easily be due to chance. In fact the 11 patients who suffered cerebral haemorrhage during this study received a total of 94 different drugs during the two years, and there were several other significant associations. Moreover, even if the

Table 10.3 Frequency of admission to hospital or
death with a diagnosis of cerebral haemorrhage
(International List 431), according to use of aspirin.

Use of aspirin	No. with cerebral haemorrhage	Total no. in study	Relative risk*
Users	3	1 577	6.9
Non-users	8	41 540	

* Adjusted for sex and age, according to the method of
Mantel and Haenszel (1959). $p < 0.01$.

association were reproducible, it could be due to some confounding variable.
Further data are needed to test the hypothesis suggested by our observation.

Discussion

The first significant attempt to use linked records for drug monitoring was by
Friedman and his colleagues in the Kaiser-Permanente group in California
(Friedman *et al.* 1971; Friedman and Collen 1972). Unfortunately, their
project was terminated before its potential could be fully assessed; its demise
appears to have been due to reliance on expensive computer-based systems
for medical records and on special recording of data by doctors (Anello
1977). Such problems can be avoided in Britain by using sources of data
already established for other purposes. Acheson (1967) recognized the
opportunity more than 10 years ago, although he considered that the
incompleteness and illegibility of identifying data on many prescriptions
could be a serious impediment. We found that this problem could be solved
by dividing the population according to general practices and comparing the
information on each prescription with a list of the identifying particulars of
people in one practice.

Our results show that it is feasible to link prescription data with details of
hospital admissions and deaths, and to analyse the information to identify
associations between drugs and events. Expected associations between
medicines and the diseases for which they are prescribed were observed; in
addition, the method detected some known adverse reactions, as well as
revealing unexpected drug-event associations.

For the purposes of the statistical analysis, it was assumed that people who
received supplies of a drug were 'users' of that drug, and that people who did
not receive prescriptions were 'non-users'. Neither of these assumptions
would always be correct: some 'users' would not have taken their medicines,
while some 'non-users' could have received prescriptions from other doctors
or (in the case of a drug such as aspirin) they could have purchased supplies

without prescription. Such misclassification does not lead to declaration of false associations; it merely reduces the power of the study to detect real associations. This could be enhanced more easily by increasing the size of the study rather than by attempting to measure compliance with therapy.

The lists of many GPs are inflated by the names of patients who have actually left the practices (Fraser 1978). If such inflation were extensive, spurious associations between drugs and events might be produced. The associations found in this study were checked by confining the analyses to people who received at least one medicine, and who were therefore known to be in the practices at the time of the study.

Because of the relatively small scale and short duration of this feasibility study, it was decided to ignore time relationships between drugs and events in the preliminary screening analysis. Time relationships were examined individually as a second phase of investigation, and this frequently obviated the need for further inquiries: for example, an association between amitriptyline and ectopic pregnancy was revealed as due to the fact that women recovering from this condition often received psychotropic drugs, rather than indicating any effect of amitriptyline on the risk of ectopic pregnancy.

Routine analyses of time relationships would be useful in a full-scale study with long-term observation. Several approaches could be employed for studying different kinds of event (for example, accidents or appearance of cancer) and some analyses could be focused on patients who received prolonged treatment.

Since our investigation was a relatively small feasibility study, it is likely that many real drug-event associations escaped detection. Nevertheless, the analyses did reveal a considerable number of statistically significant associations. The majority of these were predictable, and a large proportion of the remainder were probably due to chance. No matter what level of probability is chosen, many statistically significant associations are bound to emerge by chance from a study in which thousands of drug-event combinations are tested. Evaluation of such associations would be easier in a long-term study: analyses could be repeated periodically and each batch of new data examined separately; associations that kept recurring would be unlikely to be due to chance.

Although this partitioning of data could aid rejection of chance associations, hypotheses generated by a system of record linkage would still need to be tested in *ad hoc* studies before it could be assumed that associations were due to drug toxicity. Even if an association were not due to chance, it could be due to the influence of some confounding variable that was not controlled in the study. The most convenient way of testing hypotheses would usually be to carry out case-control studies.

Unfortunately, the number of drug-event associations signalled would probably exceed the number that could be investigated in special studies. Friedman (1972) discussed this problem and suggested that the selection of associations for further investigation should depend upon several criteria:

the degree of certainty as to the existence of the association (as indicated by its statistical significance); the likelihood that the association is causal (based on its strength and on expert appraisal); the size of the potential public health problem; and scientific implications other than those relating to drug safety.

The approach we have described, if used on a large scale for a prolonged period, should enable detection of unsuspected drug hazards, including those occurring after a delay of months or years. It would also be capable of detecting drug-related sudden death outside hospital—a catastrophic adverse effect that would often remain undiscovered by current methods of drug monitoring. Although the monitoring of hospital admissions and deaths would be most useful for detecting serious adverse reactions, recognition of the association between digoxin and diarrhoea illustrates how reactions of only moderate severity could also be identified.

Our study was too small to yield information about the effects of drugs in pregnancy, but the procedure of linking prescription data with records of pregnancies and their outcomes was found to be practicable. A larger scheme would provide an excellent opportunity to detect adverse effects of drugs on the foetus. Moreover, it should be possible to determine whether drugs increase the risk of any maternal diseases of pregnancy, such as pre-eclampsia; hazards of this nature would probably not be detected by current methods. Details of drugs prescribed for pregnant women could also be linked with records of subsequent morbidity and mortality in their offspring. This would allow inclusion of congenital malformations not diagnosed at birth, as well as other possible delayed effects of medicines. A system of record linkage would offer the best chance of detecting such hazards as the increased risk of vaginal adenocarcinoma in daughters of women given stilboestrol during early pregnancy (Herbst, Ulfelder, and Poskanzer 1971).

Other applications of record linkage

A record linkage scheme would be useful for testing existing hypotheses as well as for generating new ones. *Ad hoc* studies could involve use of the information already collected, review of case-notes, or examination of patients. We have completed two inquiries of this kind. A study of the frequency of eye complaints and rashes noted in GPs' records of patients treated with practolol and propranolol suggested that these symptoms were relatively common among patients receiving practolol (Skegg and Doll 1977). The other study showed that patients given minor tranquillizers had an increased risk of involvement in road accidents leading to hospital admission or death (Skegg, Richards, and Doll 1979).

Apart from revealing the hazards of some drugs, a record linkage scheme would provide positive evidence for the safety of many others. It might also lead to discovery of unexpected benefits of medicines. There would be opportunities for research on prescribing (Skegg, Doll, and Perry 1977), and it is possible that providing GPs with feedback about their prescribing would lead to improved use of medicines.

Requirements for a full-scale project

Before discussing the feasibility of a full-scale scheme, it is necessary to consider the size of population needed. The population required to detect a particular drug-event association depends upon the background incidence of the event, the proportion of people using the drug, and the magnitude of the risk. Even if the population included in the present study were multiplied by 10, a record linkage scheme could detect only relatively common hazards. Let us assume, for example, that a drug used by 0.5 per cent of the population caused an increase in the risk of cancer of the colon, which has an annual incidence rate in the general population of about 25 in 100 000. Table 10.4 shows that, with a study population of 500 000, the relative risk would need to be as high as 5.1 in order to have an 80 per cent chance of detecting such a hazard within two years of its appearance.

Table 10.4 Relative risks needed to have an 80 per cent probability of declaring a statistically significant alteration (two-tailed $p \leqslant 0.05$) in the risk of colon cancer among users of a drug within two years of the emergence of the hazard*

Percentage of population treated with drug	Size of population monitored†	
	500 000	5 million
2.0	2.8	1.5
0.5	5.1	2.0

  * Poisson distributions have been assumed in calculating these requirements.

  † It has been assumed that, by the time the hazard emerged, 30 per cent of the population of 500 000 (and 20 per cent of the population of five million) would have emigrated from the study area.

Now let us assume that the drug increased the risk of a relatively uncommon tumour, such as cancer of the kidney, for which the annual incidence rate is about 5 in 100 000. It can be seen from Table 10.5 that with a study population of 500 000 people, only a very large relative risk would be identified in two years after the induction period. Smaller risks could be detected after longer periods of monitoring, or if the drug were used by a higher proportion of people. But it is clear from Tables 10.4 and 10.5 that a population of five million might well be needed to detect unwanted effects quickly.

Table 10.5 Relative risks needed to have an 80 per cent
probability of declaring a statistically significant
alteration (two-tailed $p \leqslant 0.05$) in the risk of renal
cancer among users of a drug within two years of the
emergence of the hazard*

Percentage of population treated with drug	Size of population monitored†	
	500 000	5 million
2.0	5.8	2.1
0.5	11.9	3.5

* Poisson distributions have been assumed in calculating these requirements.

† It has been assumed that, by the time the hazard emerged, 30 per cent of the population of 500 000 (and 20 per cent of the population of five million) would have emigrated from the study area.

It is not possible to stipulate the optimum population needed for a full-scale drug monitoring scheme because both the value and the cost of such a scheme would increase in proportion to its size, although in different ways; ideally one would wish to include the whole population of the country. But the fact that smaller studies would fail to discover rare effects should not be a reason for inaction; the commonest hazards are the most important. Nevertheless, we must conclude that a full-scale scheme should include a population of at least half a million, and preferably as many as five million.

The Oxford Community Health Project covers a population of only 230 000, so it would not be a suitable basis for a full-scale study. However, data about basic attributes are held for every general practice in the country by the NHS family practitioner committees or the Scottish health boards. A larger drug monitoring scheme could most conveniently be established where GPs' lists are already held on computer, as in parts of Scotland. Also, in Scotland it would be possible to obtain the records of events such as we received from the Oxford Record Linkage Study (Heasman and Clarke 1979).

Would such a scheme be acceptable to the medical profession and the public? A feature attractive to doctors is the fact that the method would not add to the work of GPs. In our feasibility study we asked doctors to write person numbers on prescriptions when convenient, but this was not essential. In a full-scale project, it would be worth supplying GPs with sheets of sticky labels like those used in hospitals, bearing the names, addresses, and person numbers of patients. The doctors could then use these labels on prescriptions, and they would also find them convenient for other documents, such as laboratory request forms. Administrative staff at practices would probably be asked to assist in identifying some prescriptions and other records that could not be identified at the monitoring centre, and the practices should be reimbursed for this small amount of work.

One possible constraint would be concern about the confidentiality of medical records held on computers. It should be noted, however, that all the data that would be included in a drug monitoring scheme are already being collected for other purposes, and that, with careful design of the system, they could be protected more securely than in conventional medical records. In the final analysis, it is for the public to decide whether it is willing to accept an extremely small risk of loss of confidentiality in order to achieve a better standard of drug safety.

A more likely constraint would be cost. The annual cost of our study was about 75 pence per person monitored. Simple extrapolation would suggest that a project involving half a million people might cost something of the order of £375 000 per annum. This is a large sum but less than one-thousandth of the national drug bill. It may also be an overestimate because some expenses would not increase in proportion to the population covered. The cost could be greatly reduced if computer-based methods were adopted at the Prescription Pricing Authority, as recommended in the Tricker report (Tricker 1977).

Conclusions

We conclude that prospective drug surveillance using record linkage would be feasible and effective. A full-scale project should entail linkage of prescriptions with records of hospital admissions, obstetric deliveries, and deaths and should cover a population of at least half a million people. We hope that such a scheme can be established.

Acknowledgements

This work was carried out in the Department of the Regius Professor of Medicine, University of Oxford. We thank the following GPs who participated: Drs R. S. Pinches, A. M. Semmence, W. R. Smith, P. H. L. Tate (Abingdon); M. G. Sheldon (Banbury); T. J. Huins, D. L. Parker, P. M. M. Pritchard (Berinsfield); G. T. Smith, R. H. Stephenson, J. M. Talbot, A. J. Tulloch (Bicester); P. A. Lawrence, D. H. Richards, R. G. Seaver, A. E. Wager (Oxford). We are indebted to the Prescription Pricing Authority for providing photocopies of prescriptions, and to Mrs B. Martin and her assistants for coding them. Most of the computer programming was done by Sue Collins, Susannah Howard, and Susan Richards. We are also grateful for the collaboration of Dr J. Perry, Dr J. A. Baldwin, and other members of the Oxford Community Health Project and Oxford Record Linkage Study. We thank Professor M. P. Vessey, Mr P. G. Smith, and Mr R. Peto for their advice. The research was funded by the Department of Health and Social Security.

References

ACHESON, E. D. (1967). *Medical record linkage*. Oxford University Press, London.

ANELLO, C. (1977). *Identification of adverse reactions to marketed drugs in the United States and the United Kingdom*. Food and Drug Administration, Rockville.

BALDWIN, J. A. (1972). The Oxford Record Linkage Study as a medical information system. *Proc. R. Soc. Med.* **65**, 237.

DOLL, R. AND PETO, R. (1976). Mortality in relation to smoking: 20 years' observations on male British doctors. *Br. Med. J.* **ii**, 1525.

DUNLEAVY, J. AND PERRY, J. (1977). The Oxford Community Health Project. Paper presented at the Oxford Drug Monitoring Conference on 23 June.

FINNEY, D. J. (1965). The design and logic of a monitor of drug use. *J. Chronic. Dis.* **18**, 77.

FRASER, R. C. (1978). The reliability and validity of the age–sex register as a population denominator in general practice. *J. R. Coll. Gen. Pract.* **28**, 283.

FRIEDMAN, G. D. (1972). Screening criteria for drug monitoring. *J. Chronic. Dis.* **25**, 11.

—— AND COLLEN, M. F. (1972). A method for monitoring adverse drug reactions. In *Proceedings of the sixth Berkeley symposium on mathematical statistics and probability* (eds. Le Cam, L. M., Neyman, J., and Scott, E. L.), Vol. VI. University of California Press, Berkeley.

——, ——, HARRIS, L. E., VAN BRUNT, E. E., AND DAVIS, L. S. (1971). Experience in monitoring drug reactions in out-patients. *JAMA* **217**, 567.

HEASMAN, M. A. AND CLARKE, J. A. (1979). Medical record linkage in Scotland. *Health Bull. (Edinb.)* **37**, 97.

HERBST, A. L., ULFELDER, H., AND POSKANZER, D. C. (1971). Adenocarcinoma of the vagina. Association of maternal stilbestrol therapy with tumor appearance in young women. *N. Engl. J. Med.* **284**, 878.

MANTEL, N. AND HAENSZEL, W. (1959). Statistical aspects of the analysis of data from retrospective studies of disease. *J. Natl. Cancer Inst.* **22**, 719.

MOE, G. K. AND FARAH, A. E. (1975). Digitalis and allied cardiac glycosides. In *The pharmacological basis of therapeutics* (eds. Goodman, L. S. and Gilman, A.), Macmillan, New York.

PRESCOTT, L. F. (1975). Antipyretic analgesics. In *Meyler's side effects of drugs* (ed. Dukes, M. N. G.), Vol. VIII. Excerpta Medica, Amsterdam.

SKEGG, D. C. G. AND DOLL, R. (1977). Frequency of eye complaints and rashes among patients receiving practolol and propranolol. *Lancet* **ii**, 475.

——, —— AND PERRY, J. (1977). Use of medicines in general practice. *Br. Med. J.* **i**, 1561.

——, RICHARDS, S. M., AND DOLL, R. (1979). Minor tranquillizers and road accidents. *Br. Med. J.* **i**, 917.

——, ——, AND —— (1981). Assessment of the 'E' book as a tool for drug monitoring. *J. Epidemiol. Community Health* **35**, 32.

TRICKER, R. I. (1977). *Report of the inquiry into the Prescription Pricing Authority*. HMSO, London.

WITHERING, W. (1785). *An account of the foxglove and some of its medical uses: with practical remarks on dropsy, and other diseases*. Robinson, London.

11

Maternal anti-nauseants and clefts of lip and palate

J. GOLDING, S. VIVIAN, AND

THE LATE J. A. BALDWIN

1 Data on drug prescriptions were obtained from the general practitioners of 196 women who had had infants with clefts of lip or palate and those of 407 control women, matched for age, parity, social class, and year of delivery.
2 There was no excess of index women who had presented with nausea or vomiting.
3 There was a significant excess (12 cases, nine controls, $p < 0.02$) of women who had been prescribed Debendox (the 3-constituent, or pre-1976, formulation of Bendectin) in early pregnancy.
4 This result was not thought to be conclusive evidence of a teratogenic effect but caution in prescribing is advised pending more extensive studies.

Introduction

The possibility of anti-nauseants being teratogenic is raised from time to time. Such fears are often based initially only on one or more case reports. The probability that such suspicions are spurious is large because ingestion of anti-nauseants in the first trimester is fairly common and appears to have been rising rapidly.

Most attempts to assess the teratogenicity of various compounds have involved retrospective questioning of the mother after delivery. An important defect of such methods is that the mother who has borne a defective infant is much more likely to recall previous drug usage because she is more highly motivated and probably has spent more time already mentally recalling the events of early pregnancy. The present study was undertaken to avoid this bias although, as we shall discuss later, it probably introduces biases in a different direction.

Methods

The Oxford Record Linkage Study has obtained information on all inpatient discharges from hospitals in a defined area since 1963 (Acheson 1967; Baldwin 1972). Data abstracted include identifying information (e.g. full

Reprinted from *Human Toxicology* 2, 63–73 (1983).

name, birth surname, sex, date and place of birth) together with details of diagnoses and operations. In the case of maternities further detailed information is collected on social background, abnormalities of pregnancy and delivery, and of the infant. For domiciliary maternities within the area, midwives forward their notes to the study, from which trained clerks abstract the relevant details.

In addition to these data the Office of Population Censuses and Surveys sends to the study photocopies of livebirth and stillbirth certificates of all infants delivered to women resident in the area, regardless of where they delivered, together with death certificates of all persons resident in the area, regardless of where they died.

The present study is concerned with infants delivered between 1965 and 1974 to women resident in the area consisting of Oxfordshire and West Berkshire. The area has a total population of about 800 000 with some 14 000 births per year.

Index cases were those with clefts of lip and/or palate, whether or not other defects were present. Case ascertainment was based on the following sources: (i) diagnostic information concerning the infants for both domiciliary and hospital deliveries; (ii) diagnostic details on stillbirth and death certificates; (iii) details obtained from general hospital admissions (generally for repair of the cleft); (iv) details obtained from a malformation register kept by the Record Linkage Study; (v) cases recorded by the M.R.C. Population Genetics Research Unit.

For each index case two control deliveries were chosen matched for year of delivery, maternal age, parity, and social class. The current general practitioners (GPs) of index and control mothers still resident in the area were identified from the Oxfordshire and Berkshire Family Practitioner Committee. The N.H.S. Central Register at Southport helped trace the current GPs of women who had left the area.

Each GP was sent a letter requesting assistance, and given the names and dates of birth of women who were currently on his list. He was not told which was a case and which a control, but if he so desired he should have been able to make this identification. For each woman he was given two dates, the first being two calendar months prior to the date of the last menstrual period (LMP) and the second five calendar months after the relevant LMP. In cases where the woman was uncertain of this date, an estimate was made from the clinical details available. The GP was requested to write down only what was written in his notes concerning the period between the two dates. He was asked for details of the dates on which there had been contact with the woman, the reasons for such contact, and the drugs prescribed.

In analysing the results, gestation at any given date was calculated as the number of days (or completed weeks) from the first day of the last menstrual period (LMP). When the date of the LMP was uncertain, obstetric and neonatal information was used to determine a probable date. This was done prior to writing to the GP.

The investigation was undertaken to assess the possible teratogenicity of all drugs prescribed during pregnancy. It was not solely concerned with anti-nauseants, although this was one of the few groups that provided sufficient numbers for detailed analysis. One-tailed tests of significance have been used since the hypotheses being tested concerned teratogenicity and not protective effects. Nevertheless, it should be noted that the results of this study would still have been statistically significant if two-tailed tests had been employed throughout.

Results

In all, the sample consisted of 240 pregnancies resulting in children with clefts (91 clefts of lip and palate, 99 cleft palate alone, 50 cleft lip alone), and 480 pregnancies resulting in a child without a cleft. For various reasons information was not available for 44 cases and 73 controls.The overall response rate was similar for cases and controls: but it is of interest that the small group where no reply had been received from the GP (see Table 11.1) was significantly more likely to relate to cases than controls ($p < 0.05$). In all, data were available concerning 196 cases and 407 control pregnancies.

Ninety-one (15 per cent) of the 603 index and control women had presented with nausea and/or vomiting at some stage during the first five calendar months of pregnancy. The drugs prescribed for such nausea are listed in Table 11.2. A total of 14 different preparations had been prescribed, but only two (Debendox and Avomine) had been used sufficiently often for analysis). It can be seen (Table 11.3a) that there was no excess of index women presenting with nausea in the first five calendar months of pregnancy.

Table 11.1 Reasons for non-response

Reason	No. cases (%)	No. controls (%)
Refusal	0	2 (0.4)
Previous GP's writing illegible	0	1 (0.2)
Patient emigrated	7 (2.9)	12 (2.5)
No trace of patient's present whereabouts	12 (5.0)	21 (4.4)
No reply from GP	12 (5.0)	9 (1.9)
Patient traced but GP had not received notes from previous GP	13 (5.4)	28 (5.8)
All non-responses	44 (18.3)	73 (15.2)
All responses	196 (81.7)	407 (84.8)
Total	240	480

Table 11.2 Drugs prescribed for nausea and vomiting in the study

Name of drug	No. of pregnancies	
	Cases	Controls
Antihistamines		
Avomine (promethazine theoclate)	10 (5)	25 (18)
Debendox (dicyclomine hydrochloride, doxylamine succinate, pyridoxine hydrochloride)	13 (12)	17 (9)
Ancoloxin (meclozine hydrochloride, pyridoxine hydrochloride)	5 (3)	9 (5)
Dramamine (dimenhydrinate)	2 (1)	1 (1)
Phenergan (promethazine hydrochloride)	1 (1)	1 (0)
Vibazine (buclizine hydrochloride)	0	1 (1)
Phenothiazines		
Stemetil (prochlorperazine maleate)	1 (1)	3 (3)
Largactil (chlorpromazine hydrochloride)	0	2 (1)
Sparine (propmazine hydrochloride)	0	1 (1)
Antacids		
Aludrox (aluminium hydroxide)	0	2 (1)
Magnesium trisilicate (magnesium carbonate, magnesium trisilicate, sodium bicarbonate)	0	2 (2)
Magnesium carbonate	0	1 (1)
Prodexin (magnesium carbonate, aluminium glycinate)	0	1 (1)
Other		
Nidoxital (benzocaine, nicotinamide, pentobarbitone sodium, pyridoxine hydrochloride)	2 (2)	3 (2)

In brackets are the numbers of women who were prescribed the drug within 69 days of their last menstrual period.

Any pathological event that influences the development of cleft lip or palate, however, must take place by the end of the first nine weeks of gestation (Fitzgerald 1978). Confining the analysis therefore to women presenting with nausea within 69 days of their LMP yields the results shown in Table 11.3b. Again there was no significant excess of index women presenting with nausea, but there was a statistically significant excess of index women treated with Debendox (chi-squared with Yates' correction $= 4.9$; d.f $= 1$; $p < 0.02$), when compared with all other pregnancies, and even when compared with all other women presenting with nausea (chi-squared $= 4.5$; d.f $= 1$; $p < 0.025$). Indeed, of the 24 index mothers presenting with nausea in the first 69 days, 50 per cent had been prescribed Debendox compared with 21 per cent of the 42 controls who had attended with nausea.

A different statistical approach (Pike and Morrow 1970) involved using the matched triads (or pairs in instances where data on one control was missing). There were no instances in which two members of a triad were

Table 11.3a Women presenting to their GP with nausea at any
time during the first five calendar months of pregnancy

Treatment for nausea*	No. index pregnancies (%)	No. control pregnancies (%)
Debendox	13 (6.6)	17 (4.2)
Avomine	10 (5.1)	25 (6.1)
Other drugs	10 (5.1)	23 (5.7)
Women presenting with nausea	31 (15.8)	60 (14.7)
Total	196	407

* Not mutually exclusive.

Table 11.3b Women presenting to their GP with nausea during
the first 69 days of pregnancy

Treatment for nausea*	No. index pregnancies (%)	No. control pregnancies (%)
Debendox	12 (6.1)	9 (2.2)**
Avomine	5 (2.5)	18 (4.4)
Other drugs	7 (3.6)	15 (3.7)
No drug	1 (0.5)	2 (0.5)
Women presenting with nausea in first 9 weeks	24 (12.2)	42 (10.3)
Total	196	407

* Not mutually exclusive.
** $p < 0.02$.

positive. In accord with the general procedure, matches resulting in the triad
(− − −) or the pair (− −) have been omitted. Also omitted are all triads or
pairs where data concerning an index case were unobtainable, which explains
the reduction in the number of positive controls from nine to eight. It can be
seen from Table 11.4 that a statistically significant effect was still apparent.

The gestations of the index women at which Debendox was first prescribed
were 5, 6, 6, 7, 7, 7, 8, 8, 8, 8, 8, 9 and 10 completed weeks giving a mean of
7.46 completed weeks with standard error 0.37. The control women were
first prescribed the drug at completed weeks 6, 6, 6, 8, 8, 8, 8, 9, 9, 9, 10,10,
11, 11, 13 and 16, giving a mean of 9.29 ± 0.62 completed weeks. The means
are statistically different from one another ($t = 2.33$; d.f $= 28$; $p < 0.025$).

A further approach considers the times for which the women were
prescribed Debendox as shown in Fig. 11.1 and described in Table 11.5.
Here the actual number of weeks for which the prescription was written were

Table 11.4 Analysis of matched pairs and triads: + indicates that the mother took Debendox in the first 69 days of pregnancy; − indicates that she did not

	Index case +	Control +
Matched triad (+ − −)	9	8
Matched pair (+ −)	3	0

Observed no. cases to have taken Debendox = 12; expected no. = 7.17; chi-squared = 4.14; $p < 0.025$.

used. Where no indication of this had been given it was assumed to be for the customary three weeks. It can be seen that there was a striking contrast between index cases and controls, the cases peaking sharply at eight to nine completed weeks gestation. Using a Kolmogorov–Smirnov 2-sample test these distributions were found to be significantly different ($p < 0.001$, two-tailed). If Debendox is indeed teratogenic, the effect would thus appear to be produced at eight or nine weeks of gestation.

Controlling for nausea and vomiting it is possible to compare the two major anti-nauseants used in this period of time. In all there was a total presumed exposure to Avomine of 120 patient-weeks, and to Debendox of 113 patient-weeks. Table 11.6 shows the distribution of presumed exposure to each drug between cases and controls. The 'expected exposure' is calculated on the null hypothesis that if a drug is unassociated with the incidence of clefts then the proportion of total exposure experienced by the index cases will be the same as the proportion of index cases in the study group. This is clearly so for Avomine, but for Debendox there was a significant excess exposure among index cases.

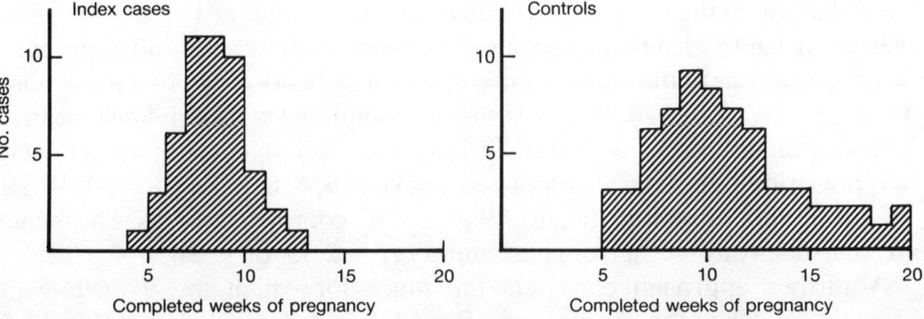

Fig. 11.1 Number of women assumed to be taking Debendox at each week of gestation.

Table 11.5 Comparison of proportion of index
and control women assumed to be taking Debendox
at each week of gestation

Completed week of gestation	Index mothers	Control mothers
5	1/96	0/407
6	3/196	3/407
7	6/196	3/407
8	11/196	6/407**
9	11/196	7/407*
10	10/196	9/407
11	4/196	8/407

* $p < 0.01$.
** $p < 0.005$.

The details of the 12 index pregnancies where the women had been prescribed Debendox are listed in Table 11.7. There was no excess of poor past obstetric outcome or of previous infants with congenital defects. One of the cases (No. 115) was a male twin of an opposite sex, non-concordant twin pair. Case 233 was probably not a Meckel syndrome since there was no evidence of renal abnormalities and Fraser and Lytwyn (1981) have stated that all cases of Meckel syndrome have cystic dysplasia of the kidneys.

Only in three of these 12 pregnancies had another drug been prescribed during the 69 days from the date of the LMP. In one of these the drugs prescribed were other anti-nauseants; in another a folic acid and vitamin preparation; and in the third an analgesic was prescribed for colic.

Discussion

The ideal design to ascertain whether a particular drug had a mild teratogenic effect would be a randomized controlled trial with women being given either

Table 11.6 Distribution of drug exposure between cases and controls

Drug	No. mothers	Total no. weeks prescribed		Significance
		Observed	Expected	
Avomine	196 cases	39	39.01	N.S.
	407 controls	81	80.99	
Debendox	196 cases	49	36.73	*$p < 0.02$
	407 controls	64	76.27	

* Chi-squared with Yates' correction.

Table 11.7 Listing of all index women prescribed Debendox in the first nine weeks of pregnancy

Case no.	Previous adverse obstetric history	Date of L.M.P.	Date Debendox prescribed	Other drugs prescribed in first nine weeks	Outcome
70	—	28.10.66	5.12.66 (5 wk)	None	Cleft lip
79	—	27.4.67	29.6.67 (9 wk)	Paracetamol (3.6.67)	Cleft palate
89	—	12.8.67	8.10.67 (8 wk)	None	Cleft lip and palate
111	—	1.7.68	1.9.68 (8 wk)	17.8.68 nidoxital 22.8.68 phenergan (both for nausea)	Cleft lip
115	—	26.9.68	18.11.68 (7 wk)	None	Cleft lip
164	1 miscarriage	7.10.70	9.12.70 (9 wk)	None	Cleft palate
175	—	14.7.71	3.9.71 (7 wk)	None	Cleft lip and palate, microcephaly, deformed left ear
177	—	21.6.71	5.8.71 (8 wk)	None	Cleft palate
225	—	12.7.73	28.8.73 (6 wk)	None	Cleft palate
228	—	24.9.73	15.11.73 (7 wk)	None	Cleft palate, sinus at back of head, bilateral talipes
233	—	6.10.73	7.12.73 (8 wk)	None	Cleft palate, encephalocele, meningomyelocele
242	—	26.2.74	23.4.74 (8 wk)	Irofol C— prescribed at same time	Cleft lip

the drug or a placebo at various (randomized) times during their pregnancies. Such a strategy would be difficult to undertake on two counts: firstly, in order to demonstrate an effect a large number of pregnant women and their GPs would have to be enrolled in the study; secondly, strict randomization might prove unacceptable to both the participants and their GPs.

An alternative design would be a prospective study, following women from the early stages of pregnancy to delivery, and following the infant thereafter, to identify any congenital defects. Again, even restricting the study to commonly prescribed drugs, the number of women required would be very large and the financial cost considerable.

Two possible strategies remain. One is to interview the mother after both she and the interviewer have discovered that her child has a congenital defect—comparing her reply with that of a control, with all the inherent biases of recall. The second is to compare available data concerning the first trimesters of such women with their controls, even though the information may be incomplete. The design of the present study probably reduced the magnitude of any teratogenic effect because it relied on the GP to report his/her own actions and prescribing habits; it can be seen from Table 11.1 that GPs of index cases appeared significantly more reluctant to reply to our request for information than were the GPs of control women. Other defects in this approach were that prescriptions were ascertained rather than consumption, and that there was no way of ascertaining what drugs mothers had bought across the counter and then consumed, whether or not they reported nausea to their GPs. Nevertheless, we suggest that this method has much to commend it if the results can be assessed in conjunction with those obtained from other (also imperfect) series.

Unlike the study of Greenberg *et al.* (1977) the cases and controls in this study were not matched on general practitioner. There were two major reasons for this: firstly it is technically not possible to do so if controls are also matched on social class, maternal age, parity, and year of delivery. For only half the cases would one matched control be feasible, and in very few would it be possible to pick two matched controls. The second point concerns the question of over-matching. For any given condition a practitioner is likely to prescribe a favourite drug at a particular point in time. To match on GP would therefore be likely to result in an inability to separate the effects of the disease from that of the drug.

In the study reported here there is the added advantage that one can compare two drugs prescribed for the same disorder. It seems evident that Avomine and Debendox did not behave in the same way as each other with respect to clefts of lip and palate. It is important to stress that data on both drugs have been collected simultaneously and in the same way, and were subjected to identical analyses. Nevertheless, they have yielded different results: Avomine appears to be unassociated with the incidence of clefts (as here defined), whereas Debendox does seem to be associated.

Analyses of the effects of anti-nauseants should take into account the possible deleterious effects of the nausea and vomiting itself. In mice, deprivation of food and water, which could well be a consequence of severe vomiting, results in a high incidence of embryos with cleft palate (Rosenzweig and Blaustein 1970). Nevertheless, large studies in Finland have shown that there appears to be no association between maternal nausea and vomiting in early pregnancy and subsequent cleft lip or palate in the child (Saxen 1975a, b). The reverse was found in a smaller study by Richards (1972) in South Wales, but the present study is in accord with Saxen's (1975a, b) findings.

The significant excess of index mothers who had been prescribed Debendox could well have been judged to be due to chance. Only one previous study, that of cases of congenital heart disease (Rothman *et al.* 1979) in the U.S.A., reported that mothers of infants with this lesion were significantly more likely than controls to have taken this drug (known as Bendectin in the U.S.A.). The relative risk computed from their figures (1.8) is somewhat less than that found in the present study (2.88).

Of the studies where the data have been collected prospectively, Smithells and Sheppard (1978) in Leeds and Liverpool found no difference in the incidence of malformations in the infants of a group of 1209 women prescribed Debendox after the 10th week. In America Milkovich and van den Berg (1976) found 3.0 per cent of malformations in 199 children whose mothers had taken Bendectin in the first 56 days compared with 3.2 per cent found in 4353 children whose mothers had had nausea and vomiting but had not been prescribed any drugs. Similar negative results were reported by Shapiro *et al.* (1977) in the U.S.A. and by Newman *et al.* (1977) in Tasmania, but in none of these studies were the numbers sufficient to demonstrate the size of effect we have shown (Vivian and Golding 1981). Harron *et al.* (1980) examined the prescription rate of Debendox in Northern Ireland over several years and compared it with the notification rate of congenital malformations (including clefts) in the province. They found an apparent inverse relationship between the rates of prescription and notification and concluded that Debendox was probably not associated with any of the malformations. Of retrospective case/control studies, Saxen (1975a, b) has identified a positive association between antihistamines and clefts of lip and/or palate in two separate studies, but no effect of Debendox on malformations was shown by Greenberg and her colleagues (1977) or by Nelson and Forfar (1977).

In view of the lack of consistent agreement, and in spite of several case reports (Paterson 1969; Donnai and Harris 1978; Mellow 1978), the findings of this study alone do not prove the case against Debendox. None the less, it is worth questioning whether prescribing of the drug for cases of mild nausea or vomiting is advisable. Hook (1974) has pointed out that nausea and vomiting appear to have an indirect beneficial effect on the fetus in that it results in a reduction in maternal smoking. Other studies have shown that nausea and vomiting in early pregnancy is associated with reduced perinatal death rates and Hook (1974) has therefore suggested that these two facts raise serious doubts about the wisdom of suppressing nausea and vomiting in early pregnancy.

Acknowledgements

We are extremely grateful to Dr C. Anello for his initial encouragement in the planning of this study and to Professor R. W. Smithells for constructive criticism. The project could not have been carried out without the assiduous assistance of Valerie Hazell, Bettine Sutton, Jean Lawrie and the general practitioners who replied to our questionnaire.

References

Acheson, E. D. (1967). *Medical Record Linkage*. Oxford University Press.

Baldwin, J. A. (1972). The Oxford Record Linkage Study as a medical information system. *Proceedings of the Royal Society of Medicine* **65**, 237–46.

Donnai, D. and Harris, R. (1978). Unusual fetal malformation after antiemetics in early pregnancy. *Brit. Med. J.* 691–2.

Fitzgerald, M. J. T. (1978). *Human Embryology*. Harper International, Maryland.

Fraser, F. C. and Lytwyn, A. (1981). Spectrum of anomalies in the Meckel syndrome. *Am. J. Med. Genet.* **9**, 67–73.

Greenberg, G., Inman, W. H., Weatherall, J. A. C., Adelstein, A. M., and Haskey, J. C. (1977). Maternal drug histories and congenital abnormalities. *Brit. Med. J.* **2**, 853–6.

Harron, D. W. G., Griffiths, K., and Shanks, R. G. (1980). Debendox and congenital malformations in Northern Ireland. *Brit. Med. J.* **281**, 1379–81.

Hook, E. B. (1974). Nausea and vomiting of pregnancy diminishes maternal smoking—is it a general feto-protective mechanism against human embryo toxins? *Teratology* A21.

Mellow, S. (1978). Fetal malformation after Debendox treatment in early pregnancy. *Brit. Med. J.* **11**, 1055–6.

Milkovich, L. and van den Berg, B. J. (1976). An evaluation of the teratogenicity of certain anti-nauseant drugs. *Am. J. Obstet. Gynecol.* **125**, 244–8.

Nelson, M. N. and Forfar, J. O. (1971). Association between drugs administered during pregnancy and congenital abnormalities of the fetus. *Brit. Med. J.* **1**, 523–7.

Newman, N. M., Correy, J. F., and Dudgeon, G. I. (1977). A survey of congenital abnormalities and drugs in private practice. *Aust. N.Z. J. Gynaecol.* **17**, 156.

Paterson, D. C. (1969). Congenital deformities. *Canad. Med. Assn. J.* **101**, 175–6.

Pike, M. C. and Morrow, R. H. (1970). Statistical analysis of patient-control studies in epidemiology. *Brit. J. Prev. Soc. Med.* **24**, 42–4.

Richards, I. D. G. (1972). A retrospective enquiry into possible teratogenic effects of drugs in pregnancy. In *Drugs and Fetal Development. Advances in Experimental Medicine and Biology* (eds. Klingber, M. A., Arbramovic, A., and Chembe, J.) **27**, 411–55. Plenum Press.

Rosenzweig, S. and Blaustein, F. M. (1970). Cleft palate in A/J mice resulting from restraint and deprivation of food and water. *Teratology* **3**, 47–52.

Rothman, K. J., Flyer, D. C., Goldblatt, A., and Kreidberg, M. B. (1979). Exogenous hormones and other drug exposures of children with congenital heart disease. *Am. J. Epid.* **109**, 433–9.

Saxen, I. (1975a). Epidemiology of cleft lip and palate. *Brit. J. Prev. Soc. Med.* **29**, 103–10.

—— (1975b). Etiological variables in oral clefts. *Proceedings Finnish Dental Soc.* Suppl. **3**, 71.

Shapiro, S., Heinonen, O. P., Siskind, V., Kaufman, D. W., Monson, R. R., and Slone, D. (1977). Antenatal exposure to doxylamine succinate and dicyclomine hydrochloride (Bendectin) in relation to congenital malformations, perinatal mortality rate, birth weight, and intelligence quotient score. *Am. J. Obstet. Gynecol.* **128**, 480–5.

Smithells, R. W. and Sheppard, S. (1978). Teratogenicity testing in humans: a method demonstrating safety of Bendectin. *Teratology* **17**, 31–6.
Vivian, S. P. and Golding, J. (1981). Debendox in early pregnancy and fetal malformation. *Brit. Med. J.* **283**, 725.

12

Isoniazid exposure in relation to cancer incidence and mortality in a cohort of tuberculosis patients

G. R. HOWE, J. LINDSAY, E. COPPOCK, AND

A. B. MILLER

Summary

41 226 isoniazid exposed and 22 811 non-exposed patients admitted to Canadian institutions for the treatment of tuberculosis from 1952 to 1960 were followed to 1973 by computerized record linkage to the National Death Index and, for 1969 to 1973, to the National Cancer Incidence Reporting System. The mortality for the total cohort compared to that expected from the mortality in the Canadian population, was greater than expected for tuberculosis, other chest diseases and certain cancer sites associated with cigarette smoking (except bladder cancer). There was, however, no difference between the isoniazid-exposed and non-exposed in mortality from all cancer and individual cancer sites, except for oesophageal cancer in males. Cancer incidence from 1969 to 1973 was also similar for the isoniazid-exposed and non-exposed groups. An elevated standardized incidence ratio was seen only for buccal cavity and pharynx in male exposed subjects. It is concluded that this study provides additional negative evidence on the carcinogenicity of isoniazid in man.

Isoniazid (INH) has been widely used since 1952 for the treatment and prophylaxis of tuberculosis (TB) because of its efficacy, low toxicity, and reasonable cost. However, it has been shown that INH can produce lung tumours, leukaemias, and lymphomas in mice (Pompe 1956; Juhasz, Balo, and Kendrey 1957; Mori and Yasuno 1959; Schwan 1961; Baincifiori and Ribacchi 1962; Toth and Shimizu 1963; Roe and Lancaster 1964; Peacock and Peacock 1966). Its possible carcinogenic potential in relation to man is therefore of great importance and a number of epidemiological studies of the drug have been conducted, mostly with negative findings (Hammond, Selikoff, and Robitzek 1967; Nyfors 1968; Ferebee 1969; Kerby *et al*. 1969; Campbell and Guilfoyle 1970; Stott *et al*. 1976; Glassroth, White, and Snider 1977). However, an elevated but non-significant relative risk was noted for females in a case control study of bladder cancer (Miller *et al*.

Reprinted from the *International Journal of Epidemiology* 8 (4), 305–12 (1979).

1978) and an excess of lung cancer from the Israeli TB Register (Steinitz 1965).

A number of the previous studies have been subject to methodological problems:

(1) INH was not used until 1952 so that if the drug is carcinogenic with a latent period typical of many carcinogens, any effect could not have manifested itself prior to the late 1960s.

(2) The sample sizes in many of the studies could not have revealed a small increase in risk. However, in view of the widespread use of the drug, even a small relative risk of cancer could result in a sizeable attributable risk.

(3) It is desirable to assemble a non-exposed comparison group with as similar characteristics as possible to the INH-exposed, in view of possible differences in cancer rates between tuberculosis patients and the general population.

Data sources and methods

Data sources

A cohort of TB patients was assembled from the Canadian National Tuberculosis Register. The register, operated by the Dominion Bureau of Statistics (now Statistics Canada) consists of records for each admission for the years 1951–60 including personal identifying data, type of admission (first or subsequent), dates of admission and discharge, diagnosis, bacillary status, whether any collapse therapy had been given, and whether INH had been given. Admissions from Ontario were excluded as individually identified cancer incidence records for that province were not available in Statistics Canada.

The mortality data consisted of computerized records derived from death certificates submitted by each province to Statistics Canada. The data used consisted of personal identifying information and date and underlying cause of death coded to the International Classification of Diseases (ICD).

The cancer incidence data were obtained from the Canadian National Cancer Reporting System, in operation since 1969. Individual records are sent by each provincial registry, with the exception of Ontario, to Statistics Canada. Records for each newly diagnosed malignant neoplasm consist of personal identifying information and nature and date of diagnosis. For the present study, data for 1969–73 was used.

Computerized record linkage

The object of any record linkage is to compare personal identifying items contained on two records and decide whether or not they refer to the same individual (Newcombe 1969a, b; Smith and Newcombe 1975). Each item, such as surname, given name, date of birth, etc. is compared and a total weight is computed, depending upon whether the items agree or disagree. The weights are based upon the frequency of occurrence of the particular value of the identifying item in the files and an estimate of how frequently particular items are misreported. The total weight obtained for the comparison of any two specific records is proportional to the odds that those two records refer to the same individual, i.e. that they are a match. Three record linkages were involved in the present study as follows.

Internal linkage of TB cohort records

It was necessary to carry out an internal linkage of the extracted TB register records to bring together those referring to multiple admissions of one individual. The identifying items used for this linkage consisted of surname, first and second given names, reporting province and institution, day, month and year of birth, sex, province and census division of residence, country of birth and, for immigrants, year of arrival in Canada. Groups of records where the weight of any link was classified as being in a 'doubtful' range, were manually inspected.

From the original 108 456 extracted records, 64 037 groups of records were formed from which single composite records were created. For 1358 records (1.3 per cent) it was not possible to decide whether the groups referred to the same individual and these were excluded.

Linkage of TB cohort and mortality records

Following the internal linkage, the composite records were linked to mortality records from 1 January 1951 to 31 December 1973. The personal identifying items used were surname, first and second given names, day, month and year of birth, sex, and province of residence. The date and cause of death were then added to those TB records which linked with a mortality record.

Linkage of TB cohort and cancer incidence records

The composite records were linked to the cancer incidence records for the years 1969–73 using the same items as for the mortality linkage. Any TB record with a definite link to a mortality record was excluded from comparison with the cancer incidence records for the years following the year of death. The date of diagnosis and the corresponding ICD code were added to any TB record which linked to an incidence record.

Methods of analysis

Since there was virtually no use of INH in Canada before 1952, the period of observation of the study was defined as 1 January 1952 to 31 December

1973. Study subjects with no record of having received INH were classified as 'non-exposed' and those with any record of having received the drug were classified as 'exposed'. It was not possible to make any estimate of the dose of the drug received nor was it possible to pin-point a specific latent period for those who received INH since the admission when the drug was first given could span a number of years; it was therefore necessary to define date of entry to the study as the admission date.

When analysing data where the end point has been determined by record linkage, certain limitations of the methodology have to be borne in mind. For any particular linkage a threshold has been chosen so that all the links having a weight greater than that value will be considered genuine matches and those below that value non-matches. If the threshold is too high, the apparent mortality or morbidity will be lower than the true rate, but if the threshold is too low, rates will be determined which are higher than the true value. In the latter case, the resulting misclassification will bring the rates for any two groups being compared towards the common population rate and thus induce a bias towards the null. If two groups being compared have linked with the same efficiency and the chosen threshold is reasonably high, the relative estimate for the two groups should be unbiased, though the absolute rates may be low. In the present study, comparisons were made between the exposed and non-exposed groups using different values for the threshold and the results obtained were very similar to those reported here. However, in comparing the rates for the TB cohort as a whole to population rates the relative rates may be biased by the choice of the threshold value since determination of the population rate is not dependent upon record linkage. Thus, for this comparison, proportional mortality was used, which should be unbiased unless the efficiency of linkage is correlated with any particular cause of death, and we have no evidence of this.

For the comparison of the exposed and non-exposed groups standardized mortality or incidence ratios (SMR or SIR) were used. Person years of observation were computed for males and females for each exposure group classified by calendar period of observation (1952–3, 1954–8, ... 1969–73), age group (0–19, 20–24, ... 80–84, 85+), and whether the individual record included day, month, and year of birth, month and year, or year only. This was necessary to control for differences in the percentage of birth day and birth month reported for the exposed and non-exposed groups. The numbers of deaths or cases of cancer occurring in the same combinations were also determined. The SMR or SIR is then computed as the ratio of the sum of the observed number of deaths or cancer cases in the exposed group to the sum of the expected number of deaths or cases. The expected numbers are derived by applying the specific calendar period, age, and birthdate information rates from the unexposed to the corresponding person years for the exposed group. Thus the SMR or SIR is a ratio of two directly standardized rates, with the standard distribution being that of the exposed group. *Standardized Proportionate Mortality Ratios* (SPMR) were computed in an analogous way for comparison of the TB cohort as a whole with Canadian population data. The denominator in the latter case was all causes of death except TB because of the great excess of TB deaths in the TB cohort. Where the total number of observations in the groups being compared is less than five, the relevant parameter is not reported, as the estimate is too unstable to be meaningful. Tests of significance and interval estimates were carried out using standard procedures (Peto and Pike 1973; Mantel and Haenszel 1959; Cutler, Schneiderman, and Greenhouse 1954; Miettinen 1976; Gart 1970; Thomas 1975).

Results

Table 12.1 shows some characteristics of the TB cohort. The non-exposed subjects tended to have earlier admissions, greater exposure to collapse therapy, were an older group and their number of admissions in the years 1952 to 1960 were fewer than the exposed. There is an excess of bacillary positive subjects among the exposed but analyses controlling for bacillary status gave virtually identical results to the analyses reported which did not control for this factor.

Comparison of TB cohort with Canadian population mortality

Table 12.2 shows SPMRs comparing the mortality of the TB cohort with that of the Canadian population for 1952–73. For males, the SPMR for all cancer is 1.00 with an upper 95 per cent confidence limit of 1.05. The proportionate mortality for three cancer sites is significantly increased (lung, buccal cavity and pharnyx, and oesophagus) and two (prostate and leukaemias and lymphomas) are decreased. For females, the SPMR for all cancer is 0.75 with an upper 95 per cent confidence limit of 0.81. Only lung cancer shows a significant elevation in SPMR while four (stomach, rectum, breast, and leukaemias and lymphomas) are decreased. Since an excess of lung cancer could be an artefact arising from misdiagnosis of lung cancer as TB, SPMRs for lung cancer were calculated for the period 1959–73, which should minimize such an effect. The values for males and females respectively are 1.14 (95 per cent confidence limits 1.02 to 1.28) and 1.26 (0.90 to 1.76). The SPMRs for chronic bronchitis and emphysema are elevated in both sexes. For males there is a significant decrease in the SPMR for heart disease and an increase for cirrhosis.

Comparison of exposed and non-exposed subjects: mortality

Table 12.3 shows SMRs for INH-exposed subjects relative to non-exposed subjects for mortality from 1952 to 1973. The exposed have an elevated mortality for TB and all causes of death except TB, but no excess for all cancer. The excess in deaths from all other causes may reflect a small residual

Table 12.1 Characteristics of tuberculosis cohort by sex and isoniazid exposure status

Characteristics	Males		Females	
	No. exposed	No. non-exposed	No. exposed	No. non-exposed
Born in Canada	19 416 (85.5%)	10 001 (86.7%)	17 098 (92.4%)	10 523 (93.3%)
Received collapse therapy	3 554 (15.6%)	2 252 (19.5%)	2 908 (15.7%)	2 560 (22.7%)
Bacillary status ever positive	14 811 (65.2%)	6 375 (55.3%)	10 412 (56.3%)	5 582 (49.5%)
Mean number of admissions (with standard deviation)	1.7 (1.2)	1.2 (0.6)	1.5 (1.0)	1.2 (0.6)
Mean age at entry (with standard deviation)	37 (18.8)	35 (18.0)	30 (16.0)	29 (14.3)
Mean year of first admission (with standard deviation)	1 955 (2.8)	1 953 (2.7)	1 955 (2.8)	1 952 (2.6)
Mean number of months for admissions during which isoniazid was given (with standard deviation)	12.9 (12.0)	—	12.4 (11.6)	—
Total number of individuals	22 720	11 530	18 506	11 281
Person years of follow-up	335 963	184 140	301 612	204 191
Average length of follow-up (years)	14.8	16.0	16.3	18.1

bias resulting from increased linkage efficiency for the exposed group. Only one individual cancer site shows a SMR that is significantly different from 1.0, namely oesophagus in the male exposed group (p two sided $= 0.032$). Two non-cancer causes of death yielded an SMR significantly different from 1.0, namely emphysema for male and cirrhosis of the liver for female exposed subjects.

Comparison of exposed and non-exposed subjects: cancer incidence

The SIRs for all sites of cancer in the TB cohort between 1969 and 1973 were 0.59 (95 per cent confidence limits 0.54 to 0.65) for males and 0.52 (0.46 to 0.58) for females. It seems unlikely that this difference is real and it is almost certainly accounted for by failure to link the records efficiently. However, relative comparisons between the exposed and non-exposed subjects should be unbiased. The SIRs comparing the exposed and non-exposed groups for various cancer sites are shown in Table 12.4. For only one site is a significantly elevated SIR seen; namely buccal cavity and pharynx for male exposed subjects.

Discussion

The data from the present study have failed to indicate any association between the use of INH and risk of cancer. In particular for the three main sites under prior consideration, namely, lung, bladder, and leukaemias and lymphomas, there is no evidence for increased incidence or mortality for those exposed to the drug as compared to those not so exposed. For some of the comparisons the number of deaths ascertained is sufficient to lead to fairly narrow confidence limits. Although it was not possible to determine a specific potential latent period for the drug, the incidence data for 1969–73 represent the experience of the exposed group between 14 and 19 years after the mean year of first admission. Thus the data cannot exclude the possibility of an effect appearing more than 20 years after first exposure. The relative rates for only two cancer sites, oesophagus and buccal cavity and pharnyx in males, are significantly elevated for mortality and incidence respectively. It seems reasonable to conclude that these results may well have been due to chance, in view of the multiple comparisons made.

The identification of INH exposed and non-exposed was based on the data on the discharge record transmitted to the Dominion Bureau of Statistics. At the time patients included in this study were admitted to institutions (1952–60), all therapy for TB would have been initiated as an in-patient. However, some of the 'non-exposed' might have received IHN post-1960 following readmission or as an out-patient. The extent of such therapy cannot be determined, but it is unlikely to have resulted in much dilution of a possible effect

Table 12.2 Standardized proportionate mortality ratios with reference to Canadian population data 1952–73

Cause of death	Male			Female		
	No. of deaths in TB group	No. of deaths in Canadian population	SPMR (95% confidence limits)	No. of deaths in TB group	No. of deaths in Canadian population	SPMR (95% confidence limits)
All causes	9 179	1 822 830		3 319	1 242 159	
TB	2 655	13 412		1 017	6 857	
All causes except TB	6 524	1 809 418		2 302	1 235 302	
All cancer	1 233	300 862	1.00 (0.94–1.05)	512	239 259	0.75*** (0.68–0.81)
Lung cancer	402	62 961	1.50*** (1.21–1.47)	50	10 683	1.50** (1.11–1.98)
Bladder cancer	25	9 050	0.73 (0.47–1.11)	2	3 318	0.38 (0.05–1.37)
Buccal cavity and pharynx cancer	69	9 037	1.95*** (1.51–2.47)	3	2 608	0.46 (0.09–1.34)
Oesophageal cancer	43	6 709	1.63** (1.18–2.20)	8	2 583	1.67 (0.72–3.29)
Stomach cancer	139	39 163	0.96 (0.81–1.14)	24	19 437	0.65* (0.41–0.97)
Colon cancer	102	29 970	0.88 (0.70–1.05)	55	33 441	0.77 (0.58–1.00)
Rectal cancer	51	15 011	0.88 (0.66–1.16)	10	9 612	0.47* (0.23–0.86)

Larynx cancer	26	3 917	1.51 (0.99–2.21)	0	563	
Bone cancer	13	2 594	1.35 (0.72–2.31)	4	1 475	1.01 (0.28–2.58)
Skin cancer	15	3 729	0.89 (0.50–1.47)	6	2 603	0.58 (0.21–1.26)
Breast cancer	0	436		116	48 693	0.65*** (0.53–0.77)
Uterine Cancer (corpus plus cervix)	0	0		70	19 077	0.95 (0.74–1.20)
Prostatic cancer	57	29 206	0.71** (0.54–0.92)	0	0	
Kidney cancer	24	7 170	0.75 (0.48–1.12)	3	3 882	0.32* (0.07–0.93)
Leukaemia and lymphomas	88	29 243	0.67*** (0.54–0.83)	41	19 888	0.64** (0.46–0.87)
Chronic bronchitis	143	15 298	2.46*** (2.08–2.92)	24	3 149	3.93*** (2.52–5.85)
Emphysema	184	13 045	3.45*** (3.00–3.99)	17	2 171	3.35*** (1.95–5.36)
Heart disease	1 721	591 686	0.76*** (0.72–0.80)	377	329 443	0.95 (0.86–1.05)
All accidents	927	195 661	0.95 (0.89–1.01)	283	73 360	1.03 (0.91–1.15)
Cirrhosis	135	18 313	1.27** (1.05–1.51)	38	9 355	0.96 (0.68–1.32)

* $p \leqslant 0.05$.
** $p \leqslant 0.01$.
*** $p \leqslant 0.001$.

Table 12.3 Standardized mortality ratios for isoniazid-exposed subjects with reference to non-exposed subjects 1952–73

Cause of death	Male			Female		
	No. of deaths in exposed	No. of deaths in non-exposed	SMR[1] (95% confidence limits)	No. of deaths in exposed	No. of deaths in non-exposed	SMR[2] (95% confidence limits)
All causes	6 211	2 967	1.09** (1.03–1.16)	2 046	1 277	1.13*** (1.05–1.21)
TB	1 736	919	1.15 (0.92–1.43)	621	400	1.31** (1.10–1.55)
All causes except TB	4 475	2 048	1.07** (1.02–1.13)	1 425	877	1.06* (1.00–1.13)
All cancer	842	391	1.01 (0.985–1.09)	293	219	0.93 (0.68–1.28)
Lung cancer	276	126	0.99 (0.88–1.12)	28	22	0.94 (0.54–2.02)
Bladder cancer	17	8	1.13 (0.37–2.50)	2	0	
Buccal cavity and pharynx cancer	52	17	1.19 (0.76–1.88)	2	1	
Oesophageal cancer	36	7	2.56* (1.02–6.45)	5	3	1.13 (0.18–6.95)
Stomach cancer	83	56	0.69 (0.39–1.22)	17	7	1.20 (0.08–17.57)
Colon cancer	70	32	1.00 (0.63–1.60)	38	17	1.39 (0.83–2.35)
Rectal cancer	34	17	0.79 (0.,20–3.08)	7	3	2.22 (0.34–9.26)

Larynx cancer	19	7	1.95 (0.57–4.26)	0	0	
Bone cancer	10	3	2.50 (0.41–11.2)	3	1	
Skin cancer	10	5	1.21 (0.27–3.24)	2	4	1.80 (0.08–7.35)
Breast cancer	0	0		59	57	0.67 (0.42–1.05)
Uterine cancer (corpus plus cervix)	0	0		45	25	2.01 (0.78–5.14)
Prostatic cancer	44	13	1.29 (0.78–2.13)	0	0	
Kidney cancer	13	11	0.56 (0.23–1.35)	1	2	
Leukaemia and lymphomas	57	31	1.08 (0.64–1.74)	21	20	0.74 (0.34–1.62)
Chronic bronchitis	103	40	1.23 (0.76–2.01)	17	7	1.26 (0.45–3.31)
Emphysema	146	38	1.63* (1.11–2.39)	11	6	1.10 (0.26–4.70)
Heart disease	1 152	568	0.97 (0.89–1.06)	235	142	0.99 (0.62–1.56)
All accidents	618	309	1.14 (0.96–1.34)	164	119	0.94 (0.05–16.8)
Cirrhosis	92	43	1.00 (0.76–1.71)	29	9	3.02* (1.29–7.10)

* $p < 0.05$.
** $p < 0.01$.
*** $p < 0.001$.

[1] The SMR is based on 335 963 person years in the exposed and 184 140 person years in the non-exposed.
[2] The SMR is based on 301 612 person years in the exposed and 204 191 person years in the non-exposed.

Table 12.4 Standardized incidence ratios for isoniazid-exposed subjects with reference to non-exposed subjects, 1969–73.

Cancer site	Male			Female		
	No. of cases in exposed	No. of cases in non-exposed	SIR[1] (95% confidence limits)	No. of cases in exposed	No. of deaths in non-exposed	SIR[2] (95% confidence limits)
All cancer	352	145	1.15 (0.92–1.45)	183	113	1.25 (0.81–1.92)
Lung cancer	58	36	0.73 (0.40–1.34)	4	4	0.98 (0.26–3.68)
Bladder cancer	12	7	0.82 (0.24–2.84)	1	0	
Buccal cavity and pharynx cancer	43	6	2.68* (1.17–6.16)	4	1	12.05 (0.001–100)
Oesophageal cancer	2	4	0.21 (0.02–2.61)	0	0	
Stomach cancer	19	9	0.67 (0.34–2.03)	1	1	
Colon cancer	29	8	3.36 (0.70–6.06)	10	11	0.66 (0.22–2.00)

Rectal cancer	21	6	1.63 (0.56–4.79)	7	3	1.51 (0.26–6.51)
Larynx cancer	6	3	0.93 (0.17–4.15)	1	0	
Bone cancer	2	0		0	0	
Skin cancer	50	31	0.97 (0.85–1.10)	23	14	1.37 (0.51–2.40)
Breast cancer	2	1		47	50	0.76 (0.48–1.19)
Uterine cancer (corpus plus cervix)	0	0		40	15	1.75 (0.86–3.54)
Prostatic cancer	26	13	0.83 (0.44–2.00)	0	0	
Kidney cancer	6	1	14.5 (0.00–100)	4	0	
Leukemia and lymphomas	21	9	1.14 (0.43–2.58)	11	2	4.03 (0.59–21.4)

* $p < 0.05, > 0.01$.

[1] The SMR is based on 84 362 person years in the exposed and 40 876 person years in the non-exposed.

[2] The SMR is based on 81 238 person years in the exposed and 47 877 person years in the non-exposed.

as any INH given would have been administered too recently for a carcinogenic effect to have been expressed.

The differences in mortality between the TB cohort as a whole and the Canadian population may be due to characteristics of the TB cohort which were not ascertainable in the present study, such as smoking habits and socioeconomic status. The sites for which significant excess mortality is seen for males; namely lung, buccal cavity and pharynx, and oesophagus, are all associated with cigarette smoking (Doll and Peto 1976). This suggests that there were more or heavier smokers in the TB cohort than the general population, although no excess of bladder cancer is seen despite the association between this disease and smoking (Howe *et al.* 1980). However, the drop in SPMR for lung cancer for both sexes for the years 1959–73 suggests that the excess of lung cancer may be due to misdiagnosis. Whatever the true cause of the observed associations, they illustrate the necessity for comparing data from INH-exposed subjects with data from non-exposed TB patients.

The study has demonstrated the utility which can be made of computerized record linkage for epidemiology. Despite a limited amount of identifying information, the linkages provided mortality and incidence data for a substantial cohort at a cost considerably less than more conventional follow-up procedures. With more identifying information, and as methodology and techniques improve, the quality and efficiency of linkage will make feasible the routine follow-up of large cohorts of exposed individuals. Such large samples will be of particular value in identifying factors which show small increases in risks or, as in the present case, for providing negative evidence on the carcinogenicity of a particular substance.

References

Baincifiori, C. and Ribacchi, R. (1962). Pulmonary tumours in mice induced by oral isoniazid and its metabolites. *Nature* **194**, 488–9.

Campbell, A. H. and Guilfoyle, P. (1970). Pulmonary tuberculosis, isoniazid, and cancer. *British Journal of Diseases of the Chest* **64**, 141–9.

Cutler, S. J., Schneiderman, M. A., and Greenhouse, S. W. (1954). Some statistical considerations in the study of cancer in industry. *American Journal of Public Health* **44**, 1159–66.

Doll, R. and Peto, R. (1976). Mortality in relation to smoking: 20 years' observations on male British doctors. *British Medical Journal* **ii**, 1525–36.

Ferebee, S. H. (1969). Controlled chemoprophylaxis trials in tuberculosis. A general review. *Advances in Tuberculosis Research* **17**, 88–106.

Gart, J. J. (1970). Point and interval estimation of the common odds ratio in the combination of 2 × 2 tables with fixed marginals. *Biometrika* **57**, 471–5.

Glassroth, J. L., White, M. C., and Snider, D. E. (1977). An assessment of the possible association of isoniazid with human cancer deaths. *American Review of Respiratory Disease* **116**, 1065–74.

Hammond, E. C. Selikoff, I. J., and Robitzek, E. H. (1967). Isoniazid therapy in relation to later occurrence of cancer in adults and infants. *British Medical Journal* ii, 792–5.

Howe, G. R., Burch, J. D., Miller, A. B., Cook, G., Esteve, J., Morrison, B., Gordon, P., Chambers, L. W., Fodor, G., and Winsor, G. M. (1980). Tobacco use, Occupation, Coffee, Various Nutrients and bladder cancer. *Journal of the National Cancer Institute* **64**, 701–13.

Juhasz, J., Balo, J., and Kendrey, G. (1957). Uber die geschwulsterzeugende Wirkung des Isonicotinsaurehydrazid (INH). *Zeitschrift für Krebsforschung* **62**, 188–96

Kerby, G. R., Rimm, A. A., Zolecki, A. A., and Stead, W. W. (1969). Cancer death rate in patients receiving prolonged isoniazid therapy. *Transactions of the 28th VA Armed Forces Pulmonary Disease Research Conference* p. 12. US Government Printing Office, Washington DC.

Mantel, N. and Haenszel, W. (1959). Statistical aspects of the analysis of data from retrospective studies of disease. *Journal of the National Cancer Institute* **22**, 719–48.

Miettinen, O. (1976). Estimability and estimation in case-referent studies. *American Journal of Epidemiology* **103**, 226–35.

Miller, C. T., Neutel, C. I., Nair, R. E., Marrett, L. D., Last, J. M., and Collins, W. E. (1978). Relative importance of risk factors in bladder carcinogenesis. *Journal of Chronic Diseases* **31**, 51–6.

Mori, K. and Yasuno, A. (1959). Preliminary note on the induction of pulmonary tumours in mice by isonicotinic acid hydrazid feeding. *Gann* **50**, 107–10.

Newcombe, H. B. (1969a). The use of medical record linkage for population and genetic studies. *Methods of Information in Medicine* **8**, 7–11.

—— (1969b). Pooled records from multiple sources for monitoring congenital anomalies. *British Journal of Preventive and Social Medicine* **23**, 226–32.

Nyfors, A. (1968). Lupus vulgaris, isoniazid, and cancer. The frequency of cancer deaths in 245 isoniazid-treated lupus vulgaris patients with an average observation period of 12 years. *Scandinavian Journal of Respiratory Diseases* **49**, 264–9.

Peacock, A. and Peacock, P. R. (1966). The results of prolonged administration of isoniazid to mice, rats, and hamsters. *British Journal of Cancer* **20**, 307.

Peto, R. and Pike, M. C. (1973). Conservatism of the Approximation $(O-E)^2/E$ in the log rank test for survival data or tumor incidence data. *Biometrics* **28**, 579–83.

Pompe, K. (1956). Einfluss von Isonicotinhydrazid auf die Lupuskarzmomentstehung. *Dermatologische Wochenschrift* **133**, 105–7.

Roe, F. J. C. and Lancaster, M. C. (1964). Natural, metallic, and other substances as carcinogens. *British Medical Bulletin* **20**, 127–33.

Schwan, S. (1961). Isonicotinic acid hydrazide (INH) as a cancerogenic agent in mice. First report. *Patologia Polska* **12**, 53.

Smith, M. E. and Newcombe, H. B. (1975). Methods for computer linkage of hospital admission-separation records into cumulative health histories. *Methods of Information in Medicine* **14**, 118–25.

Steinitz, R. (1965). Pulmonary tuberculosis and carcinoma of the lung. *American Review of Respiratory Disease* **92**, 758–66.

Stott, H., Peto, J., Stephens, R., Fox, W., Sutherland, I., Foster-Carter, A. F., Teare, H. D., and Fenning, J. (1976). An assessment of the carcinogenicity of isoniazid in patients with pulmonary tuberculosis. *Tubercle* **57**, 1–15.

THOMAS, D. G. (1975). Exact and asymptotic methods for the combination of 2 × 2 tables. *Computers and Biomedical Research* **8**, 423–46.

TOTH, B. AND SCHIMIZU, H. (1963). Lung carcinogenesis with 1-Acetyl-2-Isonicotinoylyhdrazine, the major metabolite of isoniazid. *European Journal of Cancer* **9**, 285–9.

13

Reliability of reporting by women taking part in a prospective contraceptive study

M. P. VESSEY, B. JOHNSON, AND J. DONNELLY

Summary

A prospective study is in progress at 17 family planning clinics to try to provide a balanced view of the beneficial and harmful effects of different methods of contraception. During follow-up, data about pregnancies and their outcome, hospital visits, and changes in contraceptive practices are collected from the participants in the study at routine clinic visits, by postal questionnaire, or by home visiting. Whenever a hospital admission is reported, a copy of the discharge letter or summary is obtained.

In the present investigation, the reliability of reporting of pregnancies and their outcome and hospital admissions by 1915 subjects attending two Scottish clinics was checked by comparing information obtained by the routine survey methods with maternity and hospital inpatient records maintained by the Scottish Home and Health Department. It was found that no births, miscarriages, terminations of pregnancy or admissions for sterilization had been missed in the prospective study while 90 per cent of all other hospital admissions were reported. No evidence was obtained of any important variation in the reliability of reporting between users of different methods of contraception or between different methods of collecting the follow-up information.

Introduction

In 1968, a prospective study was started in certain clinics run by the Family Planning Association to try to provide a balanced view of the beneficial and harmful effects of different methods of contraception. This investigation, which is co-ordinated by the Department of the Regius Professor of Medicine at Oxford, is now in progress at 17 clinics, and over 16 000 women are under observation.

For a clinic patient to be eligible for recruitment to the study she has to be (*a*) aged 25–39 years; (*b*) married; (*c*) a white British subject; (*d*) willing to participate; and (*e*) *either* a current user of oral contraceptives of at least five months' standing *or* a current user of the diaphragm or an intrauterine device of at least five months' standing without prior exposure to oral contraceptives. During follow-up each subject is questioned at return visits to the clinic

Reprinted from the *British Journal of Preventive and Social Medicine* **28**, 104–7 (1974).

by a doctor or a nurse, and entries are made on a special form which enable a record of pregnancies and their outcome, hospital visits, changes in contraceptive methods, and results of cervical smears taken at the clinic to be accumulated. Women who default are sent a postal version of the follow-up form and, if this is not returned, are visited in their homes for collection of the necessary information.

Data relating to hospital admissions are the principal source of information about morbidity in this study. For this reason, whenever a hospital admission is reported, contact is made with the consultant concerned and a copy of the discharge summary or letter is obtained.

From the above it will be clear that the basic source of information in this study is the woman herself. This situation immediately raises questions about the reliability of the data collected. Furthermore, because women using different methods of contraception have different patterns of clinic attendance (for example, those using oral contraceptives are usually seen every three to six months while those using the diaphragm or an intrauterine device are usually seen every six to 12 months), the possibility of bias in reporting between the three contraceptive groups must also be considered.

Fortunately, two of the family planning clinics participating in the study are situated in Scotland. This has enabled us to use the maternity and hospital inpatient records maintained by the Scottish Home and Health Department to assess the reliability of information collected about pregnancies and their outcome and hospital admissions from the women recruited to the study at these two clinics. Our findings are the subject of the present report.

Methods

Identifying information (surname, initials, and date of birth) relating to each of the 1915 subjects recruited to the prospective study at the two Scottish clinics between 1 November 1969 (the starting date of the study at these clinics) and 31 December 1971 was extracted from the index of participants. Maternities and hospital admissions experienced by these women during the years 1969–71 inclusive were then identified and listed by a search of the appropriate data files held by the Scottish Home and Health Department. Only those records where the degree of certainty of a correct match was high were considered for the present purpose. The lists were then compared with the information collected, as already described, during the course of the prospective study. During this procedure events occurring before the date of recruitment of each subject to the prospective study were, of course, ignored. It should also be noted that where two or more hospital admissions for the same disease (as defined by the three-digit International Classification of Diseases number) were listed for a subject during the follow-up period, only the first such admission was compared with the data collected during the prospective study. This restriction was imposed because all analyses of morbidity in the prospective study are based on 'first events'; repeat admissions for the same disease are therefore of no special interest.

Careful precautions were taken to ensure that confidentiality was maintained during this exercise. Firstly, the identifying information was made available by the Department of the Regius Professor of Medicine to the Scottish Home and Health Department on the strict understanding that it would be used only for the purpose already described. Secondly, the Department of the Regius Professor of Medicine undertook to make no approach to any woman about whom previously unknown information was discovered without first referring the matter back to the Scottish Home and Health Department and to the appropriate consultant. Thirdly, the entire process of comparing the data provided by the Scottish Home and Health Department with the data obtained during the course of the prospective study was carried out by just two of us, namely M.P.V., a medically qualified epidemiologist, and B.J., a senior research assistant who has signed an undertaking to maintain confidentiality in accordance with the recommendations of the Medical Research Council (1973).

Results

One livebirth and two non-obstetrical hospital admissions included on the lists provided by the Scottish Home and Health Department related to women who were lost to follow-up in the prospective study. When the remaining events were compared with the information recorded on the prospective study follow-up forms, a total of 16 hospital admissions which appeared not to have been reported were identified. It occurred to us, however, that some of these 16 admissions might represent incorrect matching rather than failures in reporting. Accordingly, a detailed investigation of these admissions was undertaken by the Scottish Home and Health Department, utilizing comprehensive identification data (including address, name of general practitioner, maiden surname, and National Health Service number) made available by the Department of the Regius Professor of Medicine. This exercise showed that five of the apparent failures in reporting were, in fact, attributable to mismatching, leaving a total of 11 genuine reporting failures.

Table 13.1 shows the main results of the present investigation, discounting the five incorrect matches. No births, miscarriages, terminations of pregnancy, or admissions for sterilization were missed in the prospective study, while 90 per cent of all other hospital admissions were reported. It is, perhaps, noteworthy that only one of the two psychiatric admissions was reported and that a slightly higher proportion of admissions for disorders of the breast or reproductive system was missed than of admissions for other disorders. There was, however, no indication of any important bias in reporting between the different contraceptive groups; thus 90, 92, and 88 per cent of the hospital admissions for reasons other than childbirth, miscarriage,

Table 13.1 Comparison of events recorded on the lists provided by the Scottish Home and Health Department with events recorded in the prospective study

Type of event	Method of contraception used by subject at entry to prospective study								
	Oral		Diaphragm		Intra uterine device		All methods		
	n^1	n^2	n^1	n^2	n^1	n^2	n^1	n^2	$100n^1/n^2$
Live birth or stillbirth	17	17	21	21	6	6	44	44	100
Miscarriage or termination	4	4	5	5	7	7	16	16	100
Admission for:									
Sterilization	14	14	5	5	7	7	26	26	100
Gynaecological or breast disease	26	30	16	18	5	6	47	54	87
Psychiatric disease	0	1	1	1	0	0	1	2	50
Any other disease	26	27	18	19	10	11	54	57	95
Total (excluding births, miscarriages, terminations, and sterilizations)	52	58	35	38	15	17	102	113	90

n^1 = Number of events on lists from Scottish Home and Health Department which were also recorded on the prospective study follow-up forms.

n^2 = Number of events on lists from Scottish Home and Health Department.

termination of pregnancy, or sterilization were reported in the oral, diaphragm, and intrauterine device contraceptive groups respectively.

Table 13.2 shows a simple analysis of the effect of some additional factors which we thought might be related to the reliability of reporting in the prospective study. The numbers are small, but there is a suggestion of less adequate reporting by women of low social class and of very short admissions. In the present context, however, the most important finding is that the collection of data by questioning at the clinic yielded closely similar results to the collection of data by other means (almost entirely by postal questionnaire).

Table 13.3 gives details of the discharge diagnoses given by the Scottish Home and Health Department for the 11 hospital admissions which were not reported in the prospective study. It can be seen that, with the exception of the one psychiatric illness, all the admissions were for minor conditions.

We also examined the relationship between the date of hospital discharge

Table 13.2 Reliability of reporting of hospital admissions for reasons other than childbirth, miscarriage, termination of pregnancy, and sterilization in relation to clinic attended, method of follow-up, age, social class, and duration of stay.

	n^1	n^2	$100n^1/n^2$
Clinic A	61	66	92
Clinic B	41	47	87
Method of follow-up			
Clinic	64	70	91
Other	38	43	88
Age (yr.)			
25–29	36	40	90
30–34	39	43	91
35–39	27	30	90
Social class			
(RG's classification)			
I–II	47	51	92
III	51	56	91
IV–V	4	6	67
Duration of stay (days)			
1	10	12	83
2–3	40	45	89
4–7	25	27	93
8 or more	27	29	93

n^1 = Number of events on lists from Scottish Home and Health Department which were also recorded on the prospective study follow-up forms.

n^2 = Number of events on lists from Scottish Home and Health Department.

Table 13.3 Discharge diagnoses provided by the Scottish Home and Health Department for the 11 admissions which were not reported in the prospective study

ICD no.	Nature of diagnosis
298.0	Reactive depressive psychosis (1)
454.9	Varicose veins without mention of ulcer (1)
611.0	Acute mastitis not associated with lactation (1)
620.0	Chronic cervicitis (1)
621.3	Chronic erosion and ulceration of cervix (1)
626.2	Excessive menstruation (2)
626.6	Intermenstrual bleeding (1)
786.0	Pain referable to urinary system (1)
Y06.0	Investigation of genital system (1)
Y10.0	General physical examination (1)

and the date at which the patient was subsequently seen at the clinic or sent a postal questionnaire. We were unable to detect any pattern which might explain why the 11 hospital admissions were not reported.

Assessment of the accuracy of the diagnostic codes in the prospective study was not an objective of the present exercise, since copies of discharge summaries or letters were routinely obtained and coding was subsequently carried out by a doctor. It may be noted, however, that of the 102 hospital admissions for reasons other than childbirth, miscarriage, termination of pregnancy, or sterilization represented both on the Scottish Home and Health Department lists and on the prospective study follow-up forms, 72 had been allocated the same three-digit International Classification of Disease list number as the principal diagnosis in both sets of data. Of the remainder, 20 showed only minor discrepancies. Major discrepancies appeared to arise most frequently when incomplete information was available at the time that data were abstracted for the Scottish Home and Health Department. For example, one patient with a final diagnosis of carcinoma-in-situ of the cervix (list no. 234.0) was allocated the diagnosis 'investigation of the genital system' (list no. Y060). It would appear that the histological report was not available at the time of preparation of the case abstract for the Scottish Home and Health Department.

Discussion

As far as the prospective study of women using different methods of contraception is concerned, the results of the present investigation are most encouraging. All births, miscarriages, terminations of pregnancy, and sterilizations recorded by the Scottish Home and Health Department were reported by the survey participants at the two Scottish clinics, while 90 per cent of other hospital admissions were also reported. Furthermore, no evidence was obtained of any important bias in reporting with respect to method of contraception or method of collecting follow-up information.

To what extent it is possible to generalize the results of this investigation to other situations is uncertain, because the women taking part in the prospective study represent a selected group of volunteers, highly motivated towards family planning. It is of interest, however, to compare the findings with those obtained in a special study undertaken by the United States Health Interview Survey (National Center for Health Statistics 1965). In this investigation, 879 people of both sexes, aged 18 years or more, who were known to have been recent hospital inpatients, were interviewed in their homes and asked about hospital discharges in the preceding year. It was found that 92 per cent of hospital discharges experienced by the men and 90 per cent experienced by women (deliveries excluded) were reported at the home interview. There was little relationship between under-reporting and age among those who were under 65 years old, while the accuracy of reporting increased with family income and educational success. Under-reporting was strongly associated with length of stay in hospital; thus for stays of only one day 72 per cent were correctly reported while for stays of eight or more days the corresponding figure was 93 per cent. With regard to the nature of the diagnosis, it was concluded that conditions which were embarrassing or 'threatening' were most likely to be under-reported—for example, only 68 per cent of psychiatric disorders and 79 per cent of disorders of the female breast and genital system were reported. Under-reporting also showed a positive relationship with the number of weeks between discharge and interview. Up to 40 weeks, however, the association was not strong; thereafter it increased sharply, probably because the respondents failed to report an appreciable number of hospital admissions of which they were aware because they were uncertain whether or not they had occurred within the one years recall period.

In general, our findings agree closely with those obtained by the United States Health Interview Survey. The two most notable differences are, firstly, that disorders of the breast and reproductive system were reported almost as reliably as other conditions in our study and, secondly, that we were unable to detect any relationship between under-reporting and the interval between discharge from hospital and questioning. The first of these differences is not difficult to explain since the participants in our prospective study would not be expected to be as embarrassed by the occurrence of disorders of the breast or reproductive system as women in the general population. The second may be related to the fact that whenever survey participants are questioned at the clinic or sent a postal questionnaire they are requested to report hospital admissions which *may* have been within the period of recall as well as those about which there is no doubt in their minds. In this way, confusion about admissions occurring around the time of the start of the period of recall, which complicated the interpretation of the Health Interview Survey, seems to have been avoided.

Acknowledgements

We express our thanks to the Medical Research Council for financial support; to Mrs D. Collinge and Mrs S. Jeffreys for careful work in the day-to-day running of the prospective study; to the research assistants, doctors, and nurses working in the family planning clinics who have made great efforts to collect reliable data; to the staff of the Scottish Office Computer Service for producing the listings of maternities and hospital admissions; and to Dr M. A. Heasman for advice and encouragement.

References

Medical Research Council (1973). Responsibility in the use of medical information for research. *Brit. med. J.* **1**, 213.

National Center for Health Statistics (1965). *Reporting of hospitalization in the Health Interview Survey*. US Department of Health, Education, and Welfare, Washington D.C.

14

Familial tendencies in diseases of children

H. B. NEWCOMBE

The clustering of cases of disease within family groups has long been of interest to geneticists, but the empirical data which they have gathered concerning the strengths of these associations have had to do largely with conditions that were thought in advance to be either hereditary or at least partially so. Epidemiologists, who might be expected to be interested in the strengths of familial predispositions, other than those due to heredity, have perhaps lacked the specific motivation of the geneticists and have certainly been hampered as well by a lack of appropriate methods for quantitative study of the problem on any substantial scale.

Conventional sampling techniques, for example, tend to be unsuitable for studies of recurrences of similar events of ill health among different members of the same families, and are especially so where the diseases in question are rare. Although this particular obstacle to investigation of the familial component of ill health might be avoided by the use of routine records of illnesses and of family compositions covering whole populations, such records have not in the past been organized in a manner that would facilitate quantitative studies of the special risks to relatives of affected individuals.

Methods are now available, however, for obtaining family histories of procreation and of mortality and morbidity from the routine records of vital and health events, throughout whole registration areas (Kennedy 1961, 1962; Kennedy *et al.* 1965; Newcombe and Kennedy 1962; Newcombe *et al.* 1959). Such methods have been used in the present study to measure the strengths of the family associations of a wide range of ailments, including many that are caused mainly by environmental factors or for which there is currently little evidence of an hereditary component in the aetiology.

Records and methods

The routine records employed in this study relate to children born in the Canadian province of British Columbia over the six-year period, 1953–8, a total of 213 392 live births and 2403 stillbirths. Of the liveborn children, 3987 had appeared in the exceptionally well organized British Columbia

Reprinted from *Br. J. Prev. Soc. Med.* **20**, 49–57 (1966).

Register of Handicapped Children and Adults by mid-1961 and were still alive, 5708 were registered as dead by the end of 1960 but not as handicapped, and a further 729 appeared in both registers. A total of 12827 children were thus affected by one or more of these kinds of registered events of misfortune, over the periods covered by the files, and 202 968 liveborn children were not so affected. Punchcards for the indexing of names and statistical purposes are produced routinely for all live births, stillbirths, deaths, and registered handicaps. These cards, or in some instances modified versions of them, were used for the present study.

The stillbirths considered in this chapter refer to foetal deaths after 28 weeks of pregnancy; these occurred at a rate of approximately 11 per thousand total births during the period under study. The rate has declined in recent years, the published figures for stillbirths per thousand live births being 11.2 for 1965 as compared with 28.1 for the period 1921 to 1925. No special check was made to assess the possible extent of under-reporting of congenital malformations as causes of stillbirths, so that the combined frequency of these conditions, and the frequencies of the particular malformations, anencephaly, hydrocephalus, and spina bifida, as given in Table 14.5, could be low by an unknown amount

The computer methods by which the vital and health records were merged and linked together into family groupings have been described in detail elsewhere (Kennedy *et al*. 1965). The resulting family master file, in magnetic tape form, contains the birth records of siblings arrayed together in order of the birth dates; interfiled between these, and following the birth records of the children to which they relate, are records of handicaps and of deaths. The arrangement is thus into family histories within which are the histories of the individual siblings.

The data-processing methods developed for this work are now exceedingly rapid, so that successive updatings of a large family master file can be carried out at a rate of one or two thousand incoming records per minute. The cost of the linkage operation is much smaller than that of producing the punchcards in the first place and in view of the continuing rapid increase in computer speeds, it is envisaged that similar procedures may in the future be widely used on a routine basis for both administrative and scientific purposes.

Extraction from the family master file of data on the recurrence of disease within sibship groups involved a number of steps:

(*a*) The production of one record per individual, summarizing in each instance the recorded history of birth, handicap, and death;

(*b*) The derivation from these of one record per family, representing a more condensed summary, which includes the histories of successive members of a sibship, with a trailing record wherever there are more individual histories in the family group than can be accommodated on a single record;

(*c*) The extraction of tabular information from family summary records selected for the presence of specified disease or groups of diseases.

These procedures are described in detail elsewhere (Smith, Schwartz, and Newcombe 1965).

The tabulations included the total numbers of earlier- and of later-born siblings of 'index' cases (i.e. of the earliest born child in each family affected by the specified condition), the number of later-born siblings who were still-born, handicapped, or dead, because of conditions other than that for which the family was selected. Such tabulations were carried out by the computer with extreme rapidity. In fact, the most time-consuming part of this and other similar studies has been the development of appropriate methods and the writing of the programmes.

Inconsistencies and changes in people's names pose special technical problems for any such study in which automatic procedures replace conventional clerical methods for matching the records pertaining to the same individuals or families by comparing names and other identifying information. Where spelling variations or changes of names are known at the time of a registration, it is customary to create supplementary punchcard records of the same events, containing the name information in its alternative forms; these are cross-referenced to one another before placing them in their respective locations in the file sequence. (The family file used for the present study was arrayed in quasi-alphabetic sequence by father's surname followed by mother's maiden surname in phonetically coded, i.e. 'Soundex' form.) Such supplementary records are necessary so that an incoming record can be linked with the rest of the family history in one or other of the alternative locations in the file sequence.

To ensure that the family group is not represented more than once in the tabulations for a particular disease cause, as a result of the presence of the supplementary records, cognizance must be taken of the cross-referencing information as indicating whether a family summary records is, or is not, the first in the file sequence representing that family. Details of the procedures used to avoid artificial inflation of the tabular information will be described in another publication (Smith *et al*. 1965); for the present, it is sufficient to say that tests carried out manually to detect failures of the cross-referencing system to avoid redundant scorings of the same family members indicate that these would lead to less than a 1 per cent inflation of the data, and not in a manner which would introduce biases.

Procedures of the above kinds will become increasingly important as attempts are made to derive family information from still larger linked files, of sizes which would render extraction of the data by human clerks even more impractical than it would have been in the present study.

Risks of similar diseases among later siblings of cases

If all kinds of registerable 'trouble' represented in the present files are treated as indistinguishable from one another, the risk of further 'trouble' among later-born siblings of an affected child is approximately 2.7 times that for the birth population as a whole (Table 14.1). The different kinds of registerable events of misfortune, namely stillbirths, handicaps, and deaths, ignoring for the moment the nature of the disease causing these events, show various degrees of familial association (Table 14.2). Stillbirths, for example, are the most likely to repeat within sibship groups, while handicaps unassociated with death are least likely to repeat; this is a somewhat unexpected finding, since many of the handicapping conditions are believed to be hereditary or partially so, whereas the extent of the contribution of hereditary causes to stillbirth is unknown.

The strengths of the familial associations have also been examined for broad categories of disease as defined by the International Classification (Table 14.3), for a number of selected diseases of possible special interest (Table 14.4), and for the different causes of stillbirth as given in the International Classification of Diseases and Causes of Death (WHO 1957) (Table 14.5).

Of the broad categories of disease (Table 14.3), some which might be regarded as largely environmental in origin, such as the infective and parasitic diseases (notably tuberculosis and poliomyelitis), diseases of the respiratory system (largely deaths from pneumonia and bronchitis), and diseases of the digestive system (largely deaths from diarrhoea), show strong familial associations, quite possibly because of substantial differences

Table 14.1 Risks of registered events of misfortune among later-born siblings of index-cases (stillbirths, handicaps, and deaths combined)

Series	Birth population	Later-born sibs of index cases
Affected children	12 827	1081*
Unaffected	202 968	6248
Affected : unaffected	1 : 15.82	1 : 5.80
Relative incidence in sibs to population	2.74	

* Event codes K, Q, R, and S = 202, 218, 627, and 34 respectively (see Table 14.2).

Table 14.2 Risks of the same events occurring among later siblings of children registered as stillborn, handicapped, or dead

Events affecting index cases	Event code	Birth population cases*	Later sibs of index cases		Relative risk (sibs to population)
			Same effect	Unaffected	
Stillbirth	K	2403	85	1070	6.71
Death only	R	5708	473	3020	5.57
Handicap and death to same child	S	729	7	409	4.77
Handicap only	Q	3987	109	1911	2.90
Weighted mean relative risk					5.08**

  * Versus 202 968 children unaffected by any of these events.
  ** Weighted geometric mean; see Woolf (1955).

between families in the quality of hygiene and child care. Deaths from accidents, poisonings, and violence are likewise strongly familial, but for a different reason; in most of the multiply-affected families, the deaths of the siblings occurred all at the same time, by drowning, fire, automobile accident, poisoning, etc.

Other broad categories, such as the allergic, endocrine, and metabolic disorders and the mental disorders of children, which are far less certain to be predominantly environmental in origin, likewise show substantial familial associations. It would be profitless, however, to speculate here concerning the relative strengths of the respective contributions from the children's genotypes and from their environments to the overall familial tendencies, because the present data do not serve to distinguish between the two sorts of cause. It is curious, however, that the congenital malformations, to which both the prenatal environments and the children's own genetic make-ups are believed to contribute substantially, exhibit the weakest measurable tendency to repeat in the same family of any of the broad categories of disease for which there are adequate data.

The relatively low position of congenital malformations, as a group, in the list of diseases with familial tendencies was one of the unexpected findings of this study. Some change in this position might perhaps be observed if allowance were made for the effects of differences in parity and parental ages; however, the relative risk value as given in Table 14.3 for congenital malformations as a group would be unlikely to be altered by as much as twofold, in view of the magnitudes of these effects as determined from previous studies (summarized in Newcombe 1965).

Of the particular causes of handicaps and deaths selected for study because they were considered to be of special interest (Table 14.4), the

Table 14.3 Risks of the same category of disease occurring among later siblings of index cases (Q = handicap only; R = death only; S = handicap and death)

Category of disease affecting index cases	Disease* code	Birth population cases**				Later sibs of index cases					Relative risk (sibs to population)
						Same disease				Unaffected	
		Q	R	S	Combined	Q	R	S	Combined		
Symptoms and ill-defined conditions	780–795	5	34	0	39	1	0	0	1	27	192.8
Allergic, endocrine system, metabolic, etc.	240–289	98	41	3	142	1	1	0	2	78	36.7
Accidents, poisonings, and violence	800–999	20	264	13	297	0	7	0	38	359	36.0
Digestive system	530–587	20	264	13	297	0	7	1	8	190	28.8
Infective and parasitic diseases	001–138	145	167	1	313	3	1	0	4	195	13.3
Mental, psychoneurotic disorders, etc.	300–326	657	7	60	724	15	0	1	16	398	11.3
Respiratory system	470–527	9	1074	0	1083	0	40	0	40	715	10.5
Diseases of early infancy	760–776	68	2936	23	3027	0	218	0	218	1399	10.4
Bones and organs of movement	720–749	514	9	38	561	7	0	0	7	269	9.4
Nervous system and sense organs	330–398	786	163	13	962	19	1	0	20	538	7.8
Neoplasms	140–239	184	25	65	274	0	1	0	1	134	5.5
Congenital malformations	750–759	1370	454	506	2330	24	6	8	8	1044	3.2
Skin and cellular tissue	690–716	32	21	1	54	0	0	0	0	39	—
Circulatory system	400–468	32	19	0	51	0	0	0	0	21	—
Genito–urinary system	590–637	20	18	2	40	0	0	0	0	23	—
Blood and blood-forming organs	290–299	10	13	1	24	0	0	0	0	9	—
Weighted mean relative risk											10.4

* International Classification of Diseases and Causes of Death (WHO 1957).

** Versus 202 968 unaffected by any registered event of misfortune.

majority show strong familial associations. Among the congenital malformations listed in the Table, 'monstrosity' has a particularly high recurrence frequency among later-born brothers and sisters of index cases ($3/24 = 12.5$ per cent; see also Table 14.11), possibly because physicians who fail to state in more specific terms the natures of the malformations tend to do so repeatedly. Malformations of the circulatory system, on the other hand, repeat only occasionally in later siblings of cases ($7/434 = 1.6$ per cent recurrence frequency). Most of the other kinds of malformations fall in between these two extremes.

Of the stillbirths due to various causes (Table 14.5), those associated with conditions affecting the mother's own health (i.e. 'acute' disease in the mother, 'other' diseases in the mother, and diseases of pregnancy and childbirth) are most likely to be followed by further stillbirths due to the same causes. Congenital malformations as a group, however, are again conspicuously low on a list arrayed in order of the strengths of the familial associations.

Non-specific associations between diseases in siblings

In addition to the tendency for particular diseases and groups of diseases to recur within families, other disease associations have been observed which are seemingly of non-specific kinds. Families selected for the presence of almost any disease or kind of event of ill health display elevated risks of stillbirths, registered handicaps, and deaths, quite apart from the kinds by which the families were distinguished in the first place. This appears to be true whether the families were chosen because of the occurrence in them of one or other of the different sorts of event, various broad categories of disease (Table 14.7), certain particular diseases of special interest (Table 14.8), or by the various causes of stillbirths (Table 14.9).

(The risks are somewhat greater for earlier-born siblings of index cases than for the later-born siblings, see Table 14.6. However, this may be due simply to the fact that the present files include handicaps and deaths at later ages where the children were born earlier in the six-year period under study.)

This non-specific elevation of the risks to families in which some sort of trouble has already occurred shows itself as an increase of approximately two-fold in the frequencies of stillbirths, handicaps, and deaths from causes other than those for which the families were selected. Because these kinds of events are relatively common, an elevation of the risks by even a small factor can result in a substantial increment in the absolute numbers of affected children. Thus, the non-specific disease associations represent, in many instances, a greater hazard to the siblings of an affected child than do repeat occurrences of the same condition (see Tables 14.10 and 14.11).

Only a small part of the non-specific associations can be attributed to more complete reporting of conditions where an earlier brother or sister has also

Table 14.4 Risks of the same selected condition among later siblings of index cases (Q = handicap only; R = death only; S = handicap and death)

Selected diseases affecting index cases	Disease code	Birth population cases*				Later sibs of index cases					Relative risk (sibs to population)
						Same disease				Unaffected	
		Q	R	S	Combined	Q	R	S	Combined		
Monstrosity	750	1	7	33	41	0	1	2	3	21	707.2
Immaturity, with other subsidiary condition	774	0	43	0	43	0	1	0	1	23	205.2
Blindness	389	51	0	0	53	2	0	1	2	42	182.4
Congenital malformation of genito–urinary system	757	90	22	28	140	0	1	2	3	57	76.3
'Other' congenital malformation of nervous system and sense organs	753	123	18	27	168	4	0	0	4	67	72.1
Strabismus	384	285	1	0	286	10	0	0	10	171	41.5
Immaturity, unqualified	766	0	1089	0	1089	0	105	0	105	476	41.1
Postnatal asphyxia and atelectasis	762	0	742	0	742	0	35	0	35	355	27.0

Congenital malformation of bone and joint	758	150	3	15	168	1	0	0	1	64	18.9
Mental deficiency	325	501	7	60	568	11	0	1	12	286	15.0
Cleft palate and hare lip	755	304	3	34	341	3	0	0	3	132	13.5
Clubfoot	748	376	0	32	408	5	0	0	5	194	12.0
'Other' and unspecified congenital malformation	759	132	42	26	200	1	0	0	1	99	10.3
Intracranial and spinal injury at birth	760	6	198	15	219	0	1	0	1	125	7.8
Congenital malformation of circulatory system	754	449	243	145	837	4	1	2	7	390	4.4
Cerebral palsy	351	207	4	8	219	0	0	0	0	84	—
Congenital malformation of digestive system	756	43	84	39	166	0	0	0	0	81	—
Spina bifida and meningocele	751	46	15	00	160	0	0	0	0	81	—
Congenital hydrocephalus	752	32	17	60	109	0	0	0	0	59	—
Epilepsy	353	72	2	1	75	0	0	0	0	50	—
Weighted mean relative risk											32.6

* Versus 202 968 unaffected.

Table 14.5 Risk of the same cause of stillbirth among later siblings of index cases.

Causes of stillbirths in index cases	Disease code	Birth population cases*	Later sibs		Relative risk (Sibs to population)
			Same cause	Unaffected	
Major causes					
Acute diseases in mother	Y 31	26	1	13	600.5
'Other' causes in mother	Y 35	32	1	18	352.4
Diseases and conditions of pregnancy	Y 32	201	5	86	58.7
Diseases of foetus and ill-defined	Y 39	595	23	226	34.7
Difficulties in labour	Y 34	182	2	98	22.8
Placental and cord conditions	Y 36	964	22	455	10.2
Congenital malformations in foetus	Y 38	262	1	125	6.2
Birth injury	Y 37	60	0	33	—
Chronic disease in mother	Y 30	50	0	11	—
Absorption of toxic substance from mother	Y 33	1	0	0	—
Weighted mean relative risk					23.0
Congenital malformations of central nervous system					
Anencephaly	Y 38.0	107	1	42	45.2
Hydrocephalus	Y 38.1	68	0	45	—
Monster	Y 38.6	6	0	4	—
Spina bifida	Y 38.2	9	0	1	—

* Versus 202 968 unaffected.

been ill. It is true that families which have come to the attention of health authorities will be more likely to be carefully screened for the presence of registerable handicaps; but this could not account for the increased frequencies of stillbirths and child deaths; because registrations of these events are virtually complete regardless of the previous health histories of the families. The effect is, in fact, greater for stillbirths and deaths than for handicaps (see footnotes to Table 14.7).

Some sort of aetiological association is implied between each of the disease causes by which the present data have been broken down and a wide range of other causes contributing to stillbirths, handicaps, and deaths. The most curious feature of these non-specific associations is that they seem rarely if ever to be absent, regardless of the index condition chosen; and furthermore, the strengths of the associations do not differ widely. This is true for families affected by such diverse categories of conditions as infections, neoplasms, and congenital malformations (Table 14.7), for particular conditions as different as blindness, cleft palate, and postnatal asphyxia (Table 14.8), and for the major causes of stillbirths whether associated with conditions in the mother, in the placenta and cord, or in the foetus (Table 14.11). Some significant heterogenity is detected by statistical test (see footnotes to Table 14.7) but the spread is not large.

A plausible partial explanation might be that many of the conditions tend to be more common in the lower income groups. This possibility could not be conveniently tested in the present study; the birth registration forms contain the occupations of the fathers but the information has not been entered in the statistical punchcards. (Unfortunately it is not at all certain that occupation will continue to be entered on Canadian birth registrations in the future.) If the suggested correlations with social factors do contribute to the observed effect, a number of different mechanisms might be involved. Environmentally determined conditions like infections and accidents are undoubtedly strongly influenced in their occurrence by family standards of child care; and the expressions of some of the partially genetic conditions (e.g. blindness, strabismus, cleft palate, spina bifida, etc.) are probably influenced by a variety of environmental factors, many of which may be correlated with social circumstances.

A further possibility exists, that troublesome genes, some of which may be relatively non-specific in their effects, are not uniformly distributed throughout the population. There is, in fact, evidence from the United Kingdom for this kind of effect as relating to mental deficiency; differential migration of persons of greater than average intelligence from rural to urban areas has given rise to regional differences in intelligence and in the incidence of mental defectives (Scottish Council for Research in Education 1953). Similar forces might perhaps contribute to other non-uniformities of genetic background throughout a population, and it is even conceivable that people living in unfavourable environments may have more than their share of unfavourable genes.

Table 14.6 Risks of other kinds of events occurring in siblings of index cases registered as stillborn, handicapped, or dead

Kinds of events affecting index cases	Event code	Birth population cases					Siblings of index cases						Unaffected	Relative risk (sibs to population)
		*Cases, excluding same kind of event					Birth rank	Cases, excluding same kind of event						
		K	Q	R	S	Combined		K	Q	R	S	Combined		
Stillbirth	K	—	3987	5708	729	10 424	Earlier	—	31	81	8	120	788	2.97
Handicap only	Q	2403	—	5708	729	8 840		30	—	82	9	121	1384	2.01
Death only	R	2403	3987	—	729	7 119		64	82	—	16	162	2597	1.81
Handicap + death	S	2403	3987	5708	—	12 098		6	8	15	—	29	258	1.89
Stillbirth	K	—	3987	5708	729	10 424	Later	—	29	76	5	110	1070	2.00
Handicap only	Q	2403	—	5708	729	8 840		31	—	93	7	131	1911	1.57
Death only	R	2403	3987	—	729	7 119		88	76	—	15	179	3020	1.69
Handicap + death	S	2403	3987	5708	—	12 098		9	9	17	—	35	409	1.44
Weighted mean relative risk														1.90

* Versus 202 968 unaffected.

Table 14.7 Risks of other kinds of diseases occurring in siblings of index cases registered as handicapped or dead according to broad categories of disease

Broad category of disease affecting index cases	Disease code	Birth population cases					Later sibs of index cases					Unaffected	Relative risk** (sibs to population)
		*Cases, excluding same kind of disease					Cases, excluding same kind of disease						
		K	Q	R	S	Combined	K	Q	R	S	Combined		
Infective and parasitic	001–138	2403	3842	5541	728	12 514	7	15	37	1	60	359	2.71
Neoplasms	140–239	2403	3803	5683	664	12 553	3	8	13	1	25	233	1.73
Allergic, endocrine, etc.	240–289	2403	3889	5667	726	12 685	2	2	12	0	16	140	1.83
Blood and blood-forming organs	290–299	2403	3977	5695	728	12 803	0	0	0	0	0	22	—
Mental, psychoneurotic, etc.	300–326	2403	3330	5701	669	12 103	5	21	38	0	64	640	1.68
Nervous system and sense organs	330–398	2403	3201	5545	716	11 865	21	22	56	4	103	856	2.06
Circulatory system	400–468	2403	3955	5689	729	12 776	2	1	5	0	8	35	3.63
Respiratory system	470–527	2403	3978	4634	729	11 744	16	40	109	3	168	1396	2.03
Digestive system	503–587	2403	3967	5444	716	12 530	4	12	46	1	63	362	2.82
Genito–urinary system	590–637	2403	3967	5690	727	12 789	0	2	3	0	5	37	2.14
Skin and cellular tissue	690–716	2403	3955	5687	728	12 773	2	2	10	0	14	62	3.58
Bones and organs of movement	720–749	2403	3473	5699	691	12 266	9	19	12	4	44	465	1.57
Congenital malformation	750–759	2403	2617	5254	223	10 497	51	41	90	4	186	1984	1.81
Certain diseases of early infancy	760–776	2403	3919	2772	706	9 800	107	67	111	18	303	2639	2.38
Symptoms and ill-defined	780–795	2403	3982	5674	729	12 788	1	1	9	0	11	39	4.48
Accidents, poisonings, violence	800–999	2403	3950	5246	726	12 325	8	14	34	2	58	56	1.69
Weighted mean relative risk***													2.12

   * Versus 202 968 unaffected.

  ** There is some statistically significant heterogeneity in the relative risks for different broad categories of disease (χ^2 for heterogeneity = 35.7; d.f. = 15; $p = 0.002$ (see Woolf 1955).

 *** The weighted mean relative risks derived separately for K, Q, R, and S are 2.91, 1.59, 2.79, and 1.58 respectively.

Table 14.8 Risks of other kinds of disease in siblings of index cases registered as handicapped or dead according to selected diseases

Selected diseases affecting index cases	Disease code	Birth population cases					Siblings of index cases					Unaffected	Relative risk (sibs to population)
		*Cases, excluding same kind of disease					Cases, excluding same kind of disease						
		K	Q	R	S	Combined	K	Q	R	S	Combined		
Mental deficiency	325	2403	3486	5701	669	12 259	4	17	30	0	51	494	1.71
Cerebral palsy	351	2403	3780	5704	721	12 608	2	7	11	3	23	141	2.63
Epilepsy	353	2403	3915	5706	728	12 752	2	0	2	0	4	72	0.88
Strabismus	384	2403	3702	5707	729	12 541	10	13	16	0	39	262	2.41
Blindness	389	2403	3936	5708	727	12 774	1	2	5	0	8	51	2.49
Clubfoot	748	2403	3611	5708	697	12 419	6	13	11	4	34	341	1.63
Monstrosity	750	2403	3986	5701	696	12 786	0	2	1	0	3	31	1.54
Spina bifida and meningocele	751	2403	3941	5693	630	12 667	7	2	4	1	14	139	1.61
Hydrocephalus	752	2403	3955	5961	669	12 718	5	4	8	1	18	124	2.32
Other congenital malformation of nervous system	753	2403	3864	5690	702	12 659	3	0	3	1	7	135	0.83

Congenital malformation of circulatory system	754	2403	3538	5465	584	11 990	18	19	37	6	80	702	1.93
Cleft palate and hare lip	755	2403	3683	5705	695	12 486	5	9	17	0	31	277	1.82
Congenital malformation of digestive system	756	2403	3944	5624	690	12 661	8	7	7	0	22	150	2.35
Congenital malformation of genito–urinary system	757	2403	3897	5686	701	12 687	2	6	3	1	12	112	1.71
Congenital malformation of bone and joint	758	2403	3837	5705	714	12 659	4	6	8	0	18	148	1.95
Other and unspecified malformation	759	2403	3855	5666	703	12 127	0	7	8	1	16	180	1.49
Intracranial and spinal injury at birth	760	2403	3981	5510	714	12 608	6	1	11	0	18	197	1.47
Postnatal asphyxia and atelectasis	762	2403	3987	5510	714	12 085	25	16	48	5	94	665	2.37
Immaturity with subsidiary condition	774	2403	3987	5665	729	12 784	0	0	9	0	8	43	2.95
Immaturity, unqualified	776	2403	3987	4619	729	11 738	36	27	63	5	131	924	2.45
Weighted mean relative risk													2.06

* Versus 202 968 unaffected.

Table 14.9 Risks of other kinds of disease in index cases registered as stillborn due to different disease causes

Causes of stillbirth affecting index cases	Disease code	Birth population cases					Siblings of index cases					Unaffected	Relative risk (sibs to population)
		*Cases, excluding same kind of disease					Cases, excluding same kind of disease						
		K	Q	R	S	Combined	K	Q	R	S	Combined		
Major causes													
Chronic disease in mother	Y 30	2353	3987	5708	729	12 777	3	0	3	0	6	18	5.30
Acute disease in mother	Y 31	2377	3987	5708	729	12 801	2	2	0	0	4	23	2.76
Disease and condition of pregnancy	Y 32	2202	3987	5708	729	12 626	6	6	7	0	19	120	2.55
Absorption of toxic substance from mother	Y 33	2402	3987	5708	729	12 826	0	0	0	0	0	0	—
Difficulties in labour	Y 34	2221	3987	5708	729	12 645	8	3	20	2	33	194	2.73
Other causes in mother	Y 35	2371	3987	5708	729	12 795	2	1	6	0	9	28	5.10
Placental and cord conditions	Y 36	1439	3987	5708	729	11 863	231	24	49	5	99	789	2.15
Birth injury	Y 37	2343	3987	5708	729	12 767	0	2	4	2	8	42	3.03
Congenital malformation of foetus	Y 38	2141	3987	5708	729	12 565	6	9	9	4	28	222	2.04
Disease of foetus and ill-defined	Y 39	1808	3987	5708	729	12 232	10	13	57	2	82	421	3.23
Weighted mean relative incidence													2.62
Congenital malformation of central nervous system													
Anencephaly	Y 38.0	2296	3987	5708	729	12 720	3	3	5	2	13	77	2.69
Hydrocephalus	Y 38.1	2335	3987	5708	729	12 759	2	3	1	2	8	70	1.82
Spina bifida	Y 38.2	2394	3987	5708	729	12 818	0	0	0	0	0	6	—
Monster	Y 38.6	2397	3987	5708	729	12 821	0	2	0	0	2	5	6.33
Weighted mean relative incidence													2.48

* Versus 202 968 unaffected.

Table 14.10 Extent of the predominance of other conditions versus the same condition among later siblings of affected children—families selected for broad categories of disease

Disease category	Code	Same condition				Other conditions				Ratio—excess other: excess same
		Ratio of cases: later sibs	No.: 10^4 later sibs	No.: 10^4 birth pop.*	Excess: 10^4 later sibs	Ratio of cases: later sibs	No.: 10^4 later sibs	No.: 10^4 birth pop.*	Excess: 10^4 later sibs	
Infective	001–138	4/232	172	15	157	33/232	1422	579	843	5.4
Neoplasm	140–239	1/142	70	13	57	7/142	493	581	—	—
Allergic	240–289	2/87	230	7	223	7/87	805	587	218	1.0
Blood	290–299	0/9	—	1	—	0/9	—	593	—	—
Mental	300–326	16/453	353	34	319	39/453	861	560	301	0.9
Nervous system	330–398	20/619	323	45	278	61/619	985	549	436	1.6
Circulatory system	400–468	0/27	—	2	—	6/27	2222	592	1630	—
Respiratory system	470–527	40/833	480	50	430	78/833	936	544	392	0.9
Digestive system	530–587	8/226	354	14	340	28/226	1239	580	659	1.9
Genito–urinary system	590–637	0/28	—	2	—	5/28	1786	592	1194	—
Skin and cellular	690–716	0/44	—	3	—	5/44	1136	591	545	—
Bones	720–749	7/302	232	26	206	26/302	861	568	293	1.4
Congenital malformation	750–759	38/1165	326	108	218	83/1165	712	486	226	1.0
Early infancy	760–776	218/1774	1229	140	1090	157/1774	885	454	431	0.4
Symptoms	780–795	1/34	294	2	292	6/34	1765	592	1173	4.0
Accidents	800–999	32/425	753	23	730	34/425	800	571	229	0.1
Combined					4340				8570	2.0

* Total birth population = 215 795. The values in these two columns should in each line add up to 594, which is the frequency, per 10^4 of the population, of handicaps plus deaths from all causes combined.

Table 14.11 Extent of the predominance of other conditions versus the same condition among later siblings of affected children—families selected for particular diseases

Disease category	Code	Same condition				Other conditions				Ratio— excess other: excess same
		Ratio of cases: later sibs	No.: 10^4 later sibs	No.: 10^4 birth pop.*	Excess: 10^4 later sibs	Ratio of cases: later sibs	No.: 10^4 later sibs	No.: 10^4 birth pop.*	Excess: 10^4 later sibs	
Mental deficiency	325	12/325	369	26	343	27/325	831	568	263	0.8
Cerebral palsy	351	0/96	—	10	—	12/96	1250	584	666	—
Epilepsy	353	0/533	—	4	—	3/53	566	590	—	—
Strabismus	384	10/207	483	13	470	26/207	1256	581	675	1.4
Blindness	389	2/50	400	3	397	6/50	1200	591	609	1.5
Clubfoot	748	5/219	228	19	209	20/219	913	575	338	1.6
Monstrosity	750	3/24	1250	2	1248	0/24	—	592	—	—
Spina bifida	751	0/89	—	7	—	8/89	899	587	312	—

Hydrocephalus	752	0/71	—	5	—	12/71	1690	589	1101	—
Other congenital malforma tion of nervous system	753	4/73	548	8	540	2/73	274	586	—	—
Congenital malformation of circulatory system	754	7/434	161	39	122	37/434	853	555	298	2.4
Cleft palate	755	3/148	203	16	187	13/148	878	578	300	1.6
Congenital malformation of digestive system	756	0/91	—	8	—	10/91	1099	586	513	—
Congenital malformation of genito–urinary system	757	3/63	476	7	469	3/63	476	587	—	—
Congenital malformation of bone	758	1/71	141	8	133	6/712	845	586	259	1.9
Other + unspecified	759	1/1108	93	9	84	8/108	741	585	156	1.9
Intracranial injury	760	1/136	74	10	64	10/136	735	584	151	2.4
Postnatal asphyxia	762	35/433	808	34	774	43/433	993	560	433	0.6
Immaturity	774	1/30	333	2	331	6/30	2000	592	1408	4.3
Immaturity, unqualified	776	105/643	1633	50	1583	62/643	964	544	420	0.3
Combined					6954				7902	1.1

*   Total birth population = 215795. The value in these two columns should in each line add up to 594, which is the frequency, per 10^4 of the birth population, of handicap plus deaths from all causes combined.

To speculate any further concerning the factors that contribute to the non-specific disease associations would be unprofitable at the present time.

Summary

The tendency for handicaps and deaths of children to be correlated, or clustered, within families has been studied, using large files of routine vital and health records from the Canadian province of British Columbia. Registrations of live births, stillbirths, and of deaths in infancy and childhood, together with records of handicapping conditions of children from an exceptionally well-organized register of handicapped children and adults, were 'linked' by computer into family groupings. The data-processing methods developed for this purpose are exceedingly rapid and permit incoming records to be entered into the appropriate family groupings at rates of 1000 to 2000 per minute. In all, 12 827 'affected' children are represented in the files, together with their unaffected brothers and sisters from a birth population of 215 795 children born in British Columbia over the six-year period 1953–8.

Almost all the conditions studied showed some tendency to repeat in the brothers and sisters of cases. For the broad categories of disease, as defined by the International Classification, the factors by which the risks from the same categories of disease were elevated among later-born siblings of 'index' cases, as compared with the birth population, ranged from 3-fold to about 200-fold. Some of the most conspicuous family groupings were not suggestive of hereditary causes, as in the case of infective diseases, respiratory diseases, and accidents. Somewhat unexpectedly, the congenital malformations as a group showed the smallest measurable tendency of any of the broad categories of disease to recur in the same families.

Among twenty selected conditions of special interest, widely different degrees of familial association were observed. 'Monstrosity', blindness, and deaths from diarrhoea were among the most likely to recur in the later-born siblings of index cases, while congenital malformations of the circulatory system showed the smallest measureable familial tendency. Among the causes of stillbirths, those associated with disease of the mother were most likely to repeat, whereas congenital malformations as a cause of stillbirth were least likely to do so.

In the families selected for the presence of various particular diseases, or broad category of disease, the brothers and sisters of the 'index' cases had, in most instances, about twice the usual risks of stillbirth, registerable handicap, and death from all other causes. These apparently non-specific disease associations give rise to about as many deaths and disabilities in the brothers and sisters of 'index' cases as do repeat occurrences of the same diseases. The effect may indicate that a wider range of diseases is more strongly correlated with social circumstances than was heretofore thought to be the case.

Acknowledgements

The computer programs by which family summary records were extracted from the linked family files were written by Miss Martha Smith; programs for the extraction of tabular information from the family summaries were written by Miss Smith and Miss Rena Schwartz. We are indebted to the British Columbia Vital Statistics Division, the Health and Welfare Division of the Dominion Bureau of Statistics, and the Computing Centre of Atomic Energy of Canada Limited for help throughout the development of the record linkage project of which this study is a part.

References

KENNEDY, J. M. (1961). *Linkage of Birth and Marriage Records using a Digital Computer*. Document No. AECL–1258, Atomic Energy of Canada Ltd., Chalk River, Ontario.

—— (1962). Use of a Digital Computer for Record Linkage. In *Use of Vital and Health Statistics for Genetic and Radiation Studies*, 155–9. UN, New York.

——, NEWCOMBE, H. B., OKAZAKI, E. A., AND SMITH, M. E. (1965). *Computer Methods for Family Linkage of Vital and Health Records*. Document No. AECL–2222, Atomic Energy of Canada Ltd., Chalk River, Ontario.

NEWCOMBE, H. B. (1965). The study of mutation and selection in human populations. *Eugen. Rev.* **57** (3), 109–25.

—— AND KENNEDY, J. M. (1962). Record linkage. Making maximum use of the discriminating power of identifying information. *Comm. Ass. comput. Mach.* **5**, 563–6.

——, ——, AXFORD, S. J., AND JAMES, A. P. (1959). Automatic linkage of vital records. *Science* **130**, 954.

Scottish Council for Research in Education (1953). *Social implications of the 1947 Scottish Mental Survey*. Publ. No. 35, University of London Press, London.

SMITH, M. E., SCHWARTZ, R. R., AND NEWCOMBE, H. B. (1965). *Computer Methods for Extracting Sibship Data from Family Groupings of Records*. Document No. AECL–2520, Atomic Energy of Canada Ltd., Chalk River, Ontario.

WOOLF, B. (1955). On estimating the relation between blood group and disease. *Ann. hum. Genet.* **19**, 251.

WHO (1957). *Manual of the International Statistical Classification of Diseases, Injuries, and Causes of Death*, based on the recommendations of the 7th Revision Conference WHO, Geneva.

15

Anencephalus in the Oxford Record Linkage Study area

J. FEDRICK*

Introduction

Anencephalus is a lethal malformation of the central nervous system. It is ideal for epidemiological study in that it is instantly recognizable and identifiable at birth and, being invariably fatal, all cases must occur either as stillbirths or early neonatal deaths. It has already been shown, though, that the fact that the lesion is recognized and can be none other than the underlying cause of death does not automatically result in a 100 per cent pick-up from any one data source (Weatherall 1969; Fedrick and Butler 1972). In any thorough investigation, a multiplicity of sources must be used.

The lesion presents many fascinating features, especially in its geographical variation. It appears to have highest incidences in Eire and Northern Ireland, closely followed by parts of the United Kingdom. Other relatively high incidence areas are found in Egypt (Stevenson *et al.* 1966), parts of the Middle East (Muffarij and Kilejian 1963) and New Zealand (Howie and Phillips 1970). From studies of, for example, the lower incidence among Irish parents living outside Ireland (Naggan and MacMahon 1967; Leck 1969), together with a lack of concordance among monozygotic twins, it has been generally accepted that the lesion has an environmental rather than genetic origin (Yen and MacMahon 1968). The possibility of a dietary factor being involved has become a popular hypothesis, with candidates including processed meat or processed peas (Knox 1972) and tea (Fedrick 1974), the case for blighted potatoes having fallen into disrepute after negative evidence from a number of studies. However, it is of little use producing hypotheses as to the aetiology of a lesion without concrete background data on the occurrence of the lesion.

There have been four comprehensive population studies of anencephalus in the British Isles: in Belfast (Elwood 1970), Liverpool (Smithells *et al.* 1964), Greater London (Carter and Evans 1973), and in the mining valleys of South Wales (Carter *et al.* 1968; Laurence *et al.* 1968a, b).

* Now Golding.

Reprinted from *Develop. Med. Child Neurol.* **18**, 643–56 (1976).

The present report is concerned with an entirely different population; that resident in the largely rural and fertile areas of Oxfordshire and Berkshire.

Material and method

The Oxford Record Linkage Study was started as an experimental project by Professor E. D. Acheson in 1962 and was expanded by its former director, the late Dr J. A. Baldwin. Its main function is to link together in a longitudinal file data concerning particular persons and, where possible, families (Acheson, 1967; Baldwin 1972).

Data collected

Information on all discharges from hospitals in the area are collected on special forms by trained clerks. Data abstracted include identifying information (e.g. full name, birth surname, sex, date and place of birth), together with details of diagnoses and operations. In the case of maternities, further detailed information is collected on social background, abnormalities of the pregnancy and delivery, and on the infant.

For domiciliary maternities within the area, a special arrangement has been made with the midwives, who forward to the study their notes, from which trained clerks abstract the relevant details.

In addition to these data, the Office of Population Censuses and Surveys send to the study photocopies of livebirth and stillbirth certificates of all infants delivered to women resident in the Record Linkage Study Area, regardless of where they were delivered, as well as the death certificates of all persons normally resident in the Area, regardless of where they died.

From 1965 to 1972 the Area covered by the maternity data consisted of the whole of Oxfordshire and that part of Berkshire within the Oxford Regional Hospital Board Area, thus covering a population of some 800 000 persons.

Method of ascertaining index series

The cases of anencephalus were ascertained using the following sources: (1) diagnostic information concerning the baby, coded on the delivery forms for both hospital and domiciliary deliveries; (2) details from stillbirth certificates; (3) details from death certificates; (4) details from a malformation register kept by the Record Linkage Study; and (5) cases recorded by the Medical Research Council's Population Genetics Research Unit of either fetuses referred for chromosome analysis or parents referred for genetic counselling. Births to all women resident in the area from 1965 to 1972 were considered.

The index series thus consisted of all anencephalics born to women resident in the area, regardless of where they were delivered. The population at risk was then taken to be the total number of births to women resident in the area over the eight-year period. For many of the present analyses, however, we had only information for the women delivered within the area, so for these analyses we considered the population at risk to consist of those women both resident and delivered in the area.

Method of ascertaining siblings

From details of the mother, such as name, birth surname, and date of birth, it was possible to identify all other pregnancies to her in the study period. Having thus ascertained the names of siblings and noted any abnormalities recognized at birth, the files of general hospital admissions were scanned to ascertain any further hospital admissions of the infants.

Definitions

The *incidence* is calculated as the rate per 1000 total births. *Parity* is the number of previous pregnancies reaching at least 28 weeks gestation or resulting in a livebirth. *Social class* is as defined by the Registrar-General for England and Wales (Great Britain 1961, 1966, 1971) according to paternal occupation as recorded on the birth certificate. The data for this item are not yet available for the whole population for 1972, therefore incidences were not able to be computed; instead, relative risks have been calculated based on population data for 1965 to 1971.

Results

Incidence

Over the eight-year period there were 177 anencephalics identified from a total population of 118 675 births. This gives an incidence of 1.49 per 1000 births, similar to the 1.41 found by Carter and Evans (1973) in Greater London, but less than the 3.95 reported for Belfast (Elwood 1970), the 3.54 of South Wales (Carter *et al*. 1968) or the 3.14 from Liverpool (Smithells *et al*. 1964).

As in most other series, there was a preponderance of females (65.5 per cent). Only 13 (7 per cent) of the cases were liveborn, and all had died within the first seven hours.

Variation within area

When looked at according to the local authority area in which the mother resided, the rate was found to vary within each area from 0 to 2.9 per 1000 births, but in only one of the 28 areas was the incidence significantly different from that found elsewhere.

In general the incidences were highest in the cities and lowest in the small towns, but the variation was small and not significant.

Secular variation

In spite of the fact that in the whole of England and Wales and in Scotland the incidence has been falling (Fedrick 1976), within the Oxford area it has risen quite dramatically (Table 15.1). In the first four years of the period the incidence averaged 1.10, but was some 73 per cent higher at 1.90 in the second half of the period ($p < 0.001$).

Table 15.1 Incidence of anencephalus by year

Year	Total no. births	No. cases of anencephalus	Incidence of anencephalus
1965	15 366	9	0.59
1966	15 327	21	1.37
1967	15 030	21	1.40
1968	14 916	16	1.07
1969	14 817	28	1.89
1970	14 776	21	1.42
1971	14 622	28	1.91
1972	13 821	33	2.39
Totals	118 675	177	1.49

χ^2 test for trend $= 15.3$, df $= 1$, $p < 0.001$.

Seasonal variation

There have been conflicting reports over the years as to whether there is a seasonal variation in the incidence of the lesion. Initial analyses revealed a significant excess of winter births in Scotland (Edwards 1958; Record 1961), in Birmingham (McKeown and Record 1951), and in Belfast (Elwood 1970), but analyses of subsequent births in the same areas revealed that the marked seasonal effect had vanished (Leck and Record 1966; Elwood and Nevin 1973; Fedrick 1976). The present data do exhibit an incidence highest in the fourth quarter of the year (Table 15.2), but the effect is not significant using either the method of Edwards (1961) or that of Hewitt *et al.* (1971).

Of far more meaning than the month of birth, however, is the month of conception—or at least the month of the last menstrual period (LMP). (This cannot be assumed to be 40 weeks earlier than the date of delivery because the gestation of an anencephalic fetus varies widely, from 28 weeks to 52 or more.)

Table 15.2 Incidence of anencephalus by month of birth

Month of birth	Total no. births*	No. cases of anencephalus	Incidence of anencephalus	
			monthly	quarterly
January	9 886	13	1.32	
February	9 598	118	1.88	1.42
March	10 888	12	1.10	
April	10 329	10	0.97	
May	10 720	11	1.03	1.19
June	10 163	16	1.57	
July	9 760	20	2.00	
August	9 760	13	1.33	1.56
September	9 748	13	1.33	
October	9 352	18	1.93	
November	9 032	17	1.88	1.86
December	9 037	16	1.77	
Totals	118 493	177	1.49	

* With recorded month of delivery.

Table 15.3 Incidence of anencephalus by month of last menstrual period (LMP)

Month of LMP	Total no. births*	No. cases of anencephalus	Incidence of anencephalus	
			monthly	quarterly
January	8 016	15	1.87	
February	7 454	12	1.87	1.65
March	8 120	12	1.48	
April	8 237	14	1.70	
May	9 248	18	1.95	1.71
June	9 376	14	1.49	
July	9 177	10	1.09	
August	9 292	8	0.86	1.21
September	8 848	15	1.70	
October	8 640	8	0.93	
November	8 334	11	1.32	1.28
December	8 882	14	1.58	
Totals	103 624	151	1.46	

* Excluding cases where month of LMP not recorded.

As can be seen in Table 15.3, for the 151 cases where the month of the LMP was recorded, there is also a variation—the incidence of anencephalus among infants conceived in the first six months of the year being some 35 per cent greater than among those conceived in the latter half of the year; but again the result is not statistically significant ($0.1 > p > 0.05$).

Paternal occupation

The occupations of the index fathers as described on the birth certificates were compared with the occupations of all fathers in the area over the years 1965 to 1971 (Table 15.4), using the codes provided by the Registrar-General (Great Britain 1966). Obviously the numbers are very small, but it is of interest that all five index fathers in the paper and printing industries were actually printers; from a population of 454 printers this gives a relative risk of 6.72 ($p < 0.001$). Note also the excessive numbers of painters and decorators ($p < 0.005$), and transport and communications workers ($p < 0.01$) who were mainly drivers of road vehicles.

Table 15.4 Paternal occupations

Occupation group	Total births 1965–71*	Index cases 1965–72	Relative risk
I Farmers, foresters, and fishermen	3 925	8	1.24
II Miners and quarrymen	81	0	—
III Gas, coke, and chemical makers	207	0	—
IV Glass and ceramics makers	70	0	—
V Furnace, forge, and foundry workers	324	2	3.76
VI Electrical and electronic workers	4 524	10	1.35
VII Engineering and allied trades	14 879	24	0.98
VIII Woodworkers	2 912	4	0.84
IX Leatherworkers	102	0	—
X Textile workers	128	0	—
XI Clothing workers	445	1	1.37
XII Food, drink, and tobacco workers	1 234	2	0.99
XIII Paper and printing workers	1 568	5	1.94
XIV Makers of other products	744	0	—
XV Construction workers	4 832	9	1.14
XVI Painters and decorators	2 178	9	2.52
XVII Drivers of stationary engines, cranes, etc.	1 142	0	—
XVIII Labourers	6 019	8	0.81
XIX Transport and communication workers	7 933	22	1.69
XX Warehousemen, storekeepers, packers, bottlers	1 927	3	0.95
XXI Clerical workers	5 502	2	0.22
XXII Sales workers	7 113	14	1.20
XXIII Service, sport, and recreation workers	4 053	3	0.45
XXIV Administrators and managers	4 217	6	0.87
XXV Professional and technical workers, artists	16 024	19	0.72
Totals	92 083	151	1.00

* Where occupation recorded (see text).

No other study has actually considered the type of paternal occupation, although various authors have shown that the incidence of anencephalus exhibits a marked trend, with lowest rates in social classes I and II and highest rates in social class V (Anderson *et al*. 1958; Edwards 1958; Butler and Alberman 1969; Elwood 1970; Fedrick 1976). The present study is no exception (Table 15.5), but in addition to this trend it can be seen that there is a marked excess of cases where the father was in the armed forces.

Table 15.5 Social class

Social class	Total births 1965–1971	Index cases 1965–72	Relative risk
I and II	24 188	25	0.61
III	47 236	85	1.07
IV and V	20 827	42	1.19
Armed forces	7 145	18	1.49
Remainder	5 458	7	0.76
Totals	104 854	177	1.00

Maternal age and parity

Authors who have compared the distributions of the ages of the mothers of anencephalics with a control population have variously found an excess of younger women (Anderson *et al*. 1958; Searle 1959; Muffarij and Kilejian 1963; Frézal *et al*. 1964; Smithells *et al*. 1964; Hay 1971), an excess of older women (Malpas 1937; Record and McKeown 1949; Coffey and Jessop 1958; Collman and Stoller 1962; Halevi 1967) or both (Ingalls *et al*. 1954; Edwards 1958; Betheras 1962; Hamersmaa 1964; Carter *et al*. 1968; Rogers 1969). Similarly, with parity there have been conflicting reports: on the one hand there are some reports of an excess of primiparae (Anderson *et al*. 1958; Hamersmaa 1964; Smithells *et al*. 1964; Williamson 1965; Czeizel and Révèsz 1970), some of an excess of high parity (Coffey and Jessop 1958; Searle 1959; Betheras 1962; Muffarij and Kilejian 1963; Frézal *et al*. 1964; Naggan 1971) and some reporting both (Ingalls *et al*. 1954; Betheras 1962; Carter *et al*. 1968; Butler and Alberman 1969).

In the present series (Tables 15.6 and 15.7) it can be seen that there is a small excess in incidence among first births, apparent in all but one age group, and an excess incidence when the mother is aged 20 to 24 years. There was no indication that high maternal age or high parity were of importance. This is in contrast with an analysis of data from the 1958 British Perinatal Mortality Survey (Fedrick 1970b), which indicated that it was infants of the very young primiparae and the older multiparae who had the highest incidences in that population.

Table 15.6 Maternal parity

Parity	Total no. births	No. index cases	Incidence	
0	41 639	65	1.56	
1	37 231	51	1.37	
2	18 679	29	1.55	
3	7 708	8	1.04	1.37
4	3 321	3	0.90	
5+	3 061	5	1.63	
All (including unknown)	118 675	177	1.49	

Table 15.7 Incidence of anencephalus by
maternal age and parity (1967–72 only)
(No. of cases in parentheses)

Maternal age	Parity		All
	0	1+	
Under 20	1.53 (9)	1.27 (2)	1.47 (11)
20–24	2.15 (31)	1.95 (28)	2.04 (59)
25–29	1.33 (11)	1.45 (28)	1.41 (39)
30–34	2.40 (5)	1.05 (11)	1.27 (16)
35+	1.44 (1)	0.91 (5)	0.96 (6)
All	1.82 (57)	1.44 (74)	1.59 (131)

Past obstetric history

In the literature there are many reports claiming that women who have borne anencephalics are likely to have had a high frequency of previous spontaneous abortion (Malpas 1937; Böök and Rayner 1950; Record and McKeown 1950a; Smithells *et al*. 1964), and previous stillbirth (Smilkstein 1962; Frézal *et al*. 1964). The present study confirms these findings (Table 15.8) and indicates that the multiparous woman who has had a prior abortion is up to 30 per cent more likely to have an anencephalic foetus, and the woman who has had a previous stillbirth is four times more likely to do so. Perhaps the most interesting case was one woman of parity 2 who had had seven abortions prior to the anencephalic.

Familial incidence

From the point of genetic counselling, obviously it is important to estimate the risk of recurrence in subsequent pregnancies. Many authors have

Table 15.8 History of previous abortion and stillbirth by parity (No. of cases in parentheses)

Parity	Percentage with previous abortion		Percentage with previous stillbirth	
	Total births 1965–72	Index cases	Total births 1965–72	Index cases
0	10.2	10.8 (7)	—	—
1	15.8	23.5 (12)	1.6	9.8 (5)
2	20.8	31.0 (9)	3.4	10.3 (3)
3+	26.0	31.2 (5)	7.0	31.2 (5)
Totals	15.8	20.5	3.2	13.5

ascertained the frequency of both anencephalus and spina bifida in the siblings of their series (Table 15.9), and have generally found that about 24 per 1000 siblings also had anencephalus and 18 per 1000 had spina bifida.

In the present study we are concerned only with those siblings born in the area between 1965 and 1972, and hence on our files. There were 53 siblings delivered prior to the index case and 62 delivered subsequently. None of these siblings were twins, and the seven cases in which the index case actually had a co-twin are discussed below.

Table 15.9 Reports in literature concerning frequency of ancencephalus and spina bifida in siblings of anencephalics

Reference	No. siblings at risk	No. with anencephalus	No. with spina bifida	Rate of defect per 1000
Penrose (1946)	48	2	2	83.3
Böök and Rayner (1950)	88	1	NS*	—
Record and McKeown (1950b)	190	1	3	21.1
MacMahon *et al.* (1953)	119	2	3	42.0
Coffey and Jessop (1958)	525	11	NS	—
Neel (1958)	80	0	0	—
Frézal *et al.* (1964)	423	5	3	18.9
Smithells *et al.* (1964)	108	2	2	37.0
Williamson (1965)	41	2	0	48.8
Carter *et al.* (1968)	708	16	13	41.0
Master-Notani *et al.* (1968)	74	0	NS	—
Smithells *et al.* (1968)	887	26	13	44.0
Yen and MacMahon (1968)	573	14	12	45.5
Czeizel and Révèsz (1970)	186	3	4	37.6
Richards *et al.* (1972)	454	18	8	57.3
Carter and Evans (1973)	754	24	18	55.7
Totals	5258	127		24.1 Anenc.
	4491		81	17.7 S.B.

* Not stated.

Altogether there were almost equal numbers of siblings of each sex, but as many as 11 of the 115 siblings had malformations of the central nervous system: there were four families which each had two cases of anencephalus, two families where a sibling was stillborn with severe spina bifida, and one where a sibling died in the neonatal period with an encephalocele. This incidence of 96 per 1000 was considerably higher than any other in the literature.

As in other series (Williamson 1965; Carter *et al*. 1968; Carter and Evans 1973), the proportion of central nervous system lesions appeared to be increased among female siblings (Table 15.10).

There were two further siblings noted as having a defect at the back of the neck: one was recorded as having a soft swelling, the other a 'buffalo hump'. There were also a stillborn infant with Down's syndrome and congenital heart disease, one infant with a large blue naevus, and one with an undescended testicle. Thus, excluding the cases with central nervous system defects, only 5 per cent of siblings had malformations; this is an unremark-able incidence.

The files were searched for any subsequent hospital admissions or deaths to those siblings surviving the neonatal period. One child born prior to the index case had died suddenly and unexpectedly at five weeks, some three weeks prior to the estimated date of the index conception. Another child had been diagnosed as having leukaemia at the age of eight weeks, four days before the LMP preceding the index conception.

Other hospital admissions of various siblings were for excision biopsy of a tumour on an ankle (found to be benign), concussion, gastro-enteritis, conjunctivitis, inguinal hernia, swallowing camphorated oil, scalds, and Perthe's disease. There were no perinatal deaths among the siblings, other than those associated with the congenital malformations.

Twins

The possibility of twinning being associated with the genesis of CNS defect has come to the fore in the hypothesis of Knox (1970) of fetus-fetus interac-

Table 15.10 Sex of the siblings of anencephalics according to the sex of index case (number of siblings with CNS defects in parentheses)

Sex of index anencephalic	No. male siblings	No. female siblings	Total
Male	17 (0)	23 (3)	40 (3)
Female	41 (4)	34 (4)	75 (8)
All	58 (4)	57 (7)	115 (11)

tion. The topic will be discussed fully in a further paper, but it should be noted here that there were seven twin births in the present study, producing an incidence of 3.1: i.e. twice as high as the incidence among single births. All were from like-sexed pairs (five female, two male), and there were no instances of concordance (Table 15.11).

Maternal blood group

Although several studies have reported no significant difference between the distribution of ABO blood groups in index mothers compared with controls (MacMahon *et al.* 1953; Muffarij and Kilejian 1963; Smithells *et al.* 1964; Carter *et al.* 1968; Czeizel and Révèsz 1970), one small study reported an excess of blood group B (Wei and Chen 1965) and the remainder have found a slight excess of O (Coffey and Jessop 1958; Penrose 1960; Collman and Stoller 1962). In the present study there is also a slight but insignificant excess of mothers of blood group O, as well as of group AB (Table 15.12). With rhesus blood group there is no appreciable difference.

Maternal religion

The religion of the overwhelming majority of women in the area is stated to be the Church of England. Numbers in other categories are too small for comment (Table 15.13).

Table 15.11 Twins in present series

Case no.	Sex of index	Sex of co-twin	Maternal age (years)	Parity	Social class	Notes
1	M	M	26	0	2	Uniovular, monoamniotic. Only two vessels in each cord. Cord joined $\frac{1}{2}$ in. before insertion in placenta. Co-twin had single transverse palmar crease; survived, but lost to follow-up (emigrated to Australia)
2	M	M	24	0	3	Uniovular, single placenta but separate amnia and chorions. Co-twin dysmature (2180 g at 40 wks); survived.
3	F	F	30	0	5	Primary infertility. Conceived on 'Volidan 21'. Co-twin 1417 g; survived.
4	F	F	20	1	5	Co-twin 1780 g; survived.
5	F	F	34	2 + 1	1	Delivered at 28 wks. Co-twin died at 12 hrs; immaturity.
6	F	F	34	6	Armed forces	Dizygotic. Co-twin 2125 g; survived
7	F	F	26	0	2	Dizygotic. Co-twin stillborn but normally developed.

Table 15.12 Incidence of anencephalus by
maternal blood group

Maternal blood group	Total births*	Index cases	Incidence
ABO group			
O	47 369	71	1.50
A	44 959	53	1.18
B	9 492	13	1.37
AB	3 274	9	2.75
Rhesus group			
+ve	87 081	124	1.42
—ve	18 013	22	1.22
Total known	105 094	146	1.39
Total unknown	6 722	15	2.23

* Within survey area only.

Illegitimacy

In early studies it has been shown that the rate of anencephalus was lower in illegitimate than legitimate births (Record and McKeown 1949; Record 1961), but reports since then have indicated that there is little difference in incidence between the two (Böök and Rayner 1950; McKeown and Record 1960; Laurence *et al*. 1968a; Butler and Alberman 1969; Horowitz and McDonald 1969; Elwood 1974). In the present series there were only nine (5.1 per cent) illegitimate births, compared with 6.4 per cent of the population as a whole; i.e. the rate among illegitimate births was slightly less than among legitimate births.

Table 15.13 Maternal religion

Religion	Total births	Cases anencephalus	Rate
Church of England	68 768	120	1.74
Roman Catholic	11 656	20	1.72
Non-conformist	7 019	9	1.28
Jewish	245	1	0.68
Other	4 151	2	
None	1 343	1	0.74
Total	93 180	159	1.70

* Where religion recorded.

Time of death in relation to sex, hydramnios, and gestation

Hydramnios has been shown to be present in approximately 50 per cent of anencephalic pregnancies (Böök and Rayner 1950; Stevenson *et al*. 1950; Stevenson 1960; Frézal *et al*. 1964; Smithells *et al*. 1964), and there has been some suggestion that it is more likely to develop in pregnancies where the anencephalic is female (Macafee 1950; MacMahon *et al*. 1953).

In the present series, hydramnios was present in 48 per cent of all pregnancies for which data were available (Table 15.14). As in previous reports, there was a tendency for hydramnios to vary with the sex of the index case, being found in 42 per cent of male pregnancies and 51 per cent of female. If the pregnancy was complicated by hydramnios, the fetus had a reduced chance of being liveborn.

Table 15.14 Hydramnios in relation to time of death and sex of fetus

Hydramnios	Stillborn		Neonatal death	
	Male	Female	Male	Female
Present	23	52	0	1
Absent	28	43	4	8
Not known	6	12	—	—
Totals	57	107	4	9

It is apparent that the pregnancy complicated by hydramnios was more likely to be delivered at early gestation (Table 15.15). Forty-five (59 per cent) of the hydramniotic pregnancies had terminated by 36 weeks, compared with 27 (32 per cent) of those without hydramnios. The sex ratio also varied with gestation: up to 38 weeks, twice as many females as males were delivered; from the 39th week, the ratio of males to females became almost one to one.

Discussion

Several features of the present study reveal differences between the incidence of anencephalus in the Oxford area and in other parts of the British Isles. There was a marked increase in the incidence of the lesion in the second half of the eight-year study period. This is in contrast with the published data from the whole of Scotland and of England and Wales (Fedrick 1976).

The other points of interest are the extremely high rate of CNS lesions in the siblings of our cases, and the high incidence of anencephalus among

Table 15.15   Gestation by sex and hydramnios

Gestation (wks)	Hydramnios present		Hydramnios absent		All*	
	M	F	M	F	M	F
≤32	4	9	2	6	8	16
33–34	5	11	3	7	10	21
35–36	4	12	4	5	9	18
37–38	3	9	4	14	8	23
39–41	3	6	10	7	13	14
42+	2	2	6	6	8	8
NS	2	4	3	6	5	16
All	23	53	32	51	61	116

* Including hydramnios not known.

twins. The latter is quite contrary to previous studies in the British Isles, where the rates in twins were approximately equal to those in single births (McKeown and Record 1960; Butler and Alberman 1969; Elwood and Nevin 1973), but is similar to that found in the Netherlands (Verstege 1971). In contrast, there were no twins in the sibships of the index cases.

Unlike most other studies there was no U-shaped incidence curve with age, rather the peak incidence was between the maternal ages of 20 and 24 years, a result also found in Hungary (Czeizel and Révèsz 1970). Nor was the distribution with parity striking, there being only a slight excess in incidence among deliveries to primiparous women.

The area covered by the study is mainly served with hard water abstracted from the River Thames. This is very similar to that supplied to Greater London, and in view of the inverse correlation between water hardness and anencephalic incidence (Fedrick 1970a; Verstege 1971), it is not surprising that the incidences in the two areas are so similar. However, an explanation for the secular increase in incidence is not immediately apparent. It is not likely to be a function of data ascertainment, as this was standard for all the years under study, and a multiplicity of data sources were employed. It will be interesting to see whether a similar result is found for any other population in the British Isles.

Summary

The incidence of anencephalus in the largely rural areas of Oxfordshire and west Berkshire from 1965 to 1972 was ascertained from the files of the Oxford Record Linkage Study as 1.49 per 1000 total births. There was little variation within the area, but there was a marked increase in incidence over the eight years of the study. There was a slight seasonal variation, with án excess of conceptions in the first two quarters of the year, a slight excess of births to primiparous women and to women aged between 20 and 24 years.

There was an increase in incidence with falling social class, with a significant excess of fathers who were printers, painters and decorators, transport drivers, or in the armed forces.

Among 115 siblings, 11 (9.6 per cent) had defects of the central nervous system, and the incidence of anencephalus among twins was twice as high as expected.

Acknowledgements

I am extremely grateful to Dr J. A. Baldwin and Professor Sir Richard Doll for constructive criticism, as well as to Mrs G. Burton, Mrs Y. Timms, and Mrs Jean Lawrie for clerical and secretarial assistance.

The information on zygosity of twin nos. 1, 2, 6 and 7 was kindly supplied by Dr Gerald Corney of the Galton Laboratory, University College, London.

References

ACHESON, E. D. (1967). *Medical Record Linkage*. London, Oxford University Press.

ANDERSON, W. J. R., BAIRD, D., AND THOMSON, A. M. (1958). Epidemiology of stillbirths and infant deaths due to congenital malformation. *Lancet* **1**, 1304.

BALDWIN, J. A. (1972). The Oxford Record Linkage Study as a medical information system. *Proc. Roy. Soc. Med.* **65**, 237.

BETHERAS, F. R. (1962). Obstetrical aspects of foetal hydrocephalus and anencephalus. *Austr. &N.Z. J. Obs. &gynae.* **2**, 31.

BÖÖK, J. A. AND RAYNER, S. (1950). A clinical and genetical study of anencephaly. *Amer. J. Hum. Gen.* **2**, 61.

BUTLER, N. R. AND ALBERMAN, E. D. (1969). *Perinatal Problems: The Second Report of the 1958 British Perinatal Mortality Survey*. E. and S. Livingstone, Edinburgh.

CARTER, C. O. AND EVANS, K. (1973). Spina bifida and anencephalus in Greater London. *J. Med. Gen.* **10**, 209.

——, DAVID, P. A., AND LAURENCE, K. M. (1968). A family study of major central nervous system malformations in South Wales. *J. Med. Gen.* **5**, 81.

COFFEY, V. P. AND JESSOP, W. J. E. (1958). A three years study of anencephaly in Dublin: a report on 181 cases. *I. J. Med. Sci.* **393**, 391.

COLLMAN, R. D. AND STOLLER, A. (1962). Relation of congenital anomalies of the central nervous system to the blood group of the mother. *Aust. &N.Z. J. Obs. & Gynae.* **2**, 38.

CZEIZEL, A. AND RÉVÈSZ, C. (1970). Major malformations of the central nervous system in Hungary. *Br. J. Prev. Soc. Med.* **24**, 205.

EDWARDS, J. H. (1958). Congenital malformations of the central nervous system in Scotland. *Br. J. Prev. Soc. Med.* **12**, 115.

—— (1961). The recognition and estimation of cyclic trends. *Ann. Hum. Gen.* **25**, 83.

ELWOOD, J. H. (1970). Anencephalus in Belfast. *Br. J. Prev. Soc. Med.* **24**, 78.

—— (1974). Anencephalus and legitimacy. *Lancet* **2**, 899.

—— AND NEVIN, N. C. (1973). Factors associated with anencephalus and spina bifida in Belfast. *Br. J. Prev. Soc. Med.* **27**, 73.

FEDRICK, J. (1970a). Anencephaly and the local water supply. *Nature* **227**, 176.

—— (1970b). Anencephalus: variation with maternal age, parity, social class, and region in England, Scotland, and Wales. *Ann. Hum. Gen.* **34**, 31.

—— (1974). Anencephalus and maternal tea-drinking: evidence for a possible association. *Proc. Roy. Soc. Med.* **67**, 356.

—— (1976). Anencephalus in Scotland 1961–1972. *Br. J. Prev. Soc. Med.* **30**, 132.

—— AND BUTLER, N. R. (1972). Accuracy of registered causes of neonatal death in 1958. *Br. J. Prev. Soc. Med.* **26**, 101.

FRÉZAL, J., KELLEY, J., GUILLEMOT, M. L., AND LAMY, M. (1964). Anencephalus in France. *Amer. J. Hum. Gen.* **16**, 336.

Great Britain: Registrar General of England and Wales (1961, 196, 1971). *Classification of Occupations*. HMSO, London.

HALEVI, H. S. (1967). Congenital malformations in Israel. *Br. J. Prev. Soc. Med.* **21**, 66.

HAMERSMAA, K. (1964). Anencefalie in Rotterdam. *Ned. Tijd. Gen.* **108**, 1109.

HAY, S. (1971). Incidence of selected congenital malformations in Iowa. *Am. J. Epid.* **94**, 572.

HEWITT, D., MILNER, J., CSIMA, A., AND PAKULA, A. (1971). On Edwards' criterion of seasonality and a non-parametric alternative. *Br. J. Prev. Soc. Med.* **25**, 174.

HOROWITZ, I. AND McDONALD, A. D. (1969). Anencephaly and spina bifida in the Province of Quebec. *Can. Med. Ass. J.* **100**, 748.

HOWIE, R. N. AND PHILLIPS, L. I. (1970). Congenital malformations in the newborn: a survey at the National Women's Hospital 1964–67. *N.Z. Med. J.* **71**, 65.

INGALLS, T. H., PUGH, T. F., AND MACMAHON, B. (1954). Incidence of anencephalus, spina bifida, and hydrocephalus related to birth rank and maternal age. *Br. J. Prev. Soc. Med.* **8**, 17.

KNOX, E. G. (1970). Fetus-fetus interaction—a model aetiology for anencephalus. *Dev. Med. &Ch. Neur.* **12**, 167.

—— (1972). Anencephalus and dietary intakes. *Br. J. Prev. Soc. Med.* **26**, 219.

LAURENCE, K. M., CARTER, C. O., AND DAVID, P. A. (1968a). Major central nervous system malformations in South Wales: I. Incidence, local variations, and geographical factors. *Br. J. Prev. Soc. Med.* **22**, 146.

——, ——, —— (1968b). Major central nervous system malformations in South Wales: II. Pregnancy factors, seasonal variation, and social class effects. *Br. J. Prev. Soc. Med.* **22**, 212.

LECK, I. (1969). Ethnic differences in the incidence of malformations following migration. *Br. J. Prev. Soc. Med.* **23**, 166.

—— AND RECORD, R. G. (1966). Seasonal incidence of anencephaly. *Br. J. Prev. Soc. Med.* **20**, 67.

MACAFEE, C. H. G. (1950). Hydramnios. *J. Obs. &Gynae. Br. Emp.* **57**, 171.

MACMAHON, B., PUGH, T. F., AND INGALLS, T. H. (1953). Anencephalus, spina bifida, and hydrocephalus: incidence related to sex, race, and season of birth and incidence in siblings. *Br. J. Prev. Soc. Med.* **7**, 211.

MALPAS, P. (1937). The incidence of human malformations and the significance of changes in the maternal environment in their causation. *J. Obs. & Gynae. Br. Emp.* **44**, 434.

Master-Notani, P., Kolah, P. J., and Sanghvi, L. D. (1968). Congenital malformations in Bombay. II. *Act. Gen. (Basel)* **18**, 193.

McKeown, T. and Record, R. G. (1951). Seasonal incidence of congenital malformations of the central nervous system. *Lancet* **1**, 192.

——, —— (1960). Malformations in a population observed for five years. In *Ciba Foundation Symposium on Congenital Malformations* (eds. Wolstenholme, G. E. W. and O'Connor, M.). Churchill, London.

Muffarij, I. K. and Kilejian, V. O. (1963). Anencephaly: an analysis of anencephalic births and a report of a case of repeated anencephaly. *Obs. & Gyn.* **22**, 657.

Naggan, L. (1971). Anencephaly and spina bifida in Israel. *Pediatrics* **47**, 577.

—— and MacMahon, B. (1967). Ethnic differences in the prevalence of anencephaly and spina bifida in Boston, Massachusetts. *N. Engl. J. Med.* **277**, 1119.

Neel, J. V. (1958). A study of major congenital defects in Japanese infants. *Amer. J. Hum. Gen.* **10**, 398.

Penrose, L. S. (1946). Familial data on 144 cases of anencephaly and spina bifida and congenital hydrocephaly. *Ann. Eug.* **13**, 73.

—— (1960). Genetical causes of malformations and the search for their origins. In *Ciba Foundation Symposium on Congenital Malformations* (eds. Wolstenholme, G. E. W. and O'Connor, M.). Churchill, London.

Record, R. G. (1961). Anencephalus in Scotland. *Br. J. Prev. Soc. Med.* **15**, 93.

—— and McKeown, T. (1949). Congenital malformations of the central nervous system: I. A survey of 90 cases. *Br. J. Soc. Med.* **3**, 183.

——, —— (1950a). Congenital malformations of the central nervous system: II. Maternal reproductive history and familial incidence. *Br. J. Soc. Med.* **4**, 26.

——, —— (1950b). Congenital malformations of the central nervous system: III. Risk of malformation in sibs of malformed individuals. *Br. J. Soc. Med.* **4**, 217.

Richards, I. D. G., McIntosh, H. J., and Sweenie, S. (1972). A genetic study of anencephaly and spina bifida in Glasgow. *Dev. Med. &Ch. Neur.* **14**, 626.

Rogers, S. C. (1969). Epidemiology of stillbirths from congenital abnormalities in England and Wales, 1961–1966. *Dev. Med. &Ch. Neur.* **11**, 617.

Searle, A. G. (1959). The incidence of anencephaly in a polytypic population. *Ann. Hum. Gen.* **23**, 279.

Smilkstein, G. (1962). A ten-year study of anencephaly. *Cal. Med.* **96**, 350.

Smithells, R. W., Chinn, E. R., and Franklin, D. (1964). Anencephaly in Liverpool. *Dev. Med. &Ch. Neur.* **6**, 231.

——, D'Arcy, E. E., and McAlister, E. F. (1968). The outcome of pregnancies before and after the birth of infants with nervous system malformations. *Dev. Med. &Ch. Neur.* (Suppl.) **15**, 6.

Stevenson, A. C. (1960). The association of hydramnios with congenital malformations. In *Ciba Foundation Symposium on Congenital Malformations* (eds. Wolstenholme, G. E. W. and O'Connor, M.). Churchill, London.

——, Johnston, H. A., Stewart, M. I. P., and Golding, D. R. (1966). Congenital malformations: a report of a study of series of consecutive births in 24 centres. *Bull. WHO* **34** (Suppl.).

Stevenson, S. S., Worcester, J., and Rice, R. G. (1950). Six hundred and seventy-seven congenitally malformed infants and associated gestational characteristics. I. General considerations. *Pediatrics* **6**, 37.

VERSTEGE, J. CH. W. (1971). *Anencephalie in Nederland 1951–1968*. Central Bureau voor Statistiek, Staatsuitgeverij (The Hague).

WEATHERALL, J. A. C. (1969). An assessment of the efficiency of notification of congenital malformations. *Med. Off.* **121**, 65.

WEI, P. Y. AND CHEN, Y. P. (1965). Congenital malformations, especially anencephalus, in Taiwan. *Am. J. Obs. &Gyn.* **91**, 870.

WILLIAMSON, E. M. (1965). Incidence and family aggregation of major congenital malformations of the central nervous system. *J. Med. Gen.* **2**, 161.

YEN, S. AND MACMAHON, B. (1968). Genetics of anencephaly and spina bifida? *Lancet* **2**, 623.

16

Families with maltreated children in north-east Wiltshire

J. E. OLIVER AND W. J. GRAHAM

Introduction

The problem of child abuse has been with us since time immemorial. In the medical profession, growing awareness of extreme violence to infants dates from 1946, when Caffey described the true significance of the association between subdural haematomas and fractures of the long bones in children (Smith and Hanson 1974). Interest was aroused in England after the publication of case reports in 1963 (Griffiths and Moynihan 1963) and the British Paediatric Association's warning memorandum in 1966, which helped define the problem and offered guidelines for treatment.

It is now recognized that the syndromes of child abuse and neglect contribute to both childhood morbidity and mortality in Western Europe and the United States (Smith 1978; Christoffel, Liu, and Stamler, 1981). It is also apparent that these syndromes often go unrecognized, undiagnosed, and underestimated. The reasons for this situation are complex, but one of the major contributory factors is the difficulty of obtaining reliable information (Oliver *et al*. 1974). Detection, even for the purpose of research, is an arduous task of fitting together pieces of information from many sources, recorded or remembered over lengthy periods. Yet there is evidence to suggest that, had the available information been linked into family and personal histories, at least some of the damage inflicted upon the children might have been prevented. This chapter provides such evidence and, in so doing, illustrates the importance of linking records at the level of clinical care.

The study population

North-east Wiltshire was the first part of Great Britain to start child maltreatment registers and, for certain families, has linked information, derived from multiple sources, for over 25 years. This information is centred on children perceived to be at risk of neglect or assault within their own families. As the relevant agencies have co-operated since the 1950s, there are family-linked data pertaining to child maltreatment over more than one generation for certain families within a defined geographical area. This presents a unique opportunity to consider child abuse over a long period.

The primary aim of this study was to ascertain the number of families with two or more generations of maltreated children and multi-agency involvement, in which the occurrence of child abuse has continued to the present day. Secondary aims included the assessment of family circumstances and the types and extent of maltreatment. This was was, in fact, a continuation of research which has been carried out since 1969 (Oliver and Taylor 1971; Oliver and Cox 1973; Oliver *et al*. 1974; Oliver, Cox, and Buchanan 1978).

All the families studied lived within the Swindon Health District, comprising the local authority districts of Thamesdown, part of Kennet, and part of north Wiltshire. The population of this area of Wiltshire has grown from 169 000 in 1961 to over 214 000 in 1981, with the town of Swindon accounting for a substantial proportion of the growth. In 1971, when the population of the area was 200 000, there were 44 800 children under twelve years of age and 3600 live births.

Families and Methods of Study

Selection criteria

The criteria for selection of the families were as follows:

(1) Each family contained at least two children (with at least one common parent) who had been resident in the area for a minimum of three months, had been born between 1960 and 1980, and had been neglected or maltreated at home *and* (unless killed) received intervention by child protection agencies to prevent further maltreatment.

(2) The index children had a mother or father (the latter living with them for at least three months) who had been neglected or maltreated in childhood and had received intervention by child protection agencies.

(3) At least two children and one parent, or one child and both parents, must each have received multi-agency support, defined in terms of three or more social support agencies, or five or more agency workers providing a measure of social support or control (see Oliver and Cox 1973).

(4) The index children were maltreated between the years of 1960 and 1980 while living within the Swindon Health District. All the children entering into the subsequent analysis were first maltreated before they were twelve years of age.

Data collection and linkage

The methods of case detection which had to be adopted were those of person and family record linkage. Ascertainment was based essentially on the principal author's clinical responsibilities and involvement with the 34 agencies which either had records or direct concern for young children at risk within the study area. Information was obtained from these agencies, from child abuse case-conferences, from the clinical care of one or more individuals (in three-quarters of the families), and also from previous research (Oliver *et al*, 1974).

Access to this highly sensitive information was heavily dependent on trust and close collaboration with the various agencies and the families. The obstacles to data collection included confidentiality, inappropriate patterns of response by professionals, parental denial or understatement of events, as well as the fact that information was unavailable, withheld, or destroyed. Some of these difficulties have been further described elsewhere (Oliver and Graham 1985).

The process of gathering and cross-referencing information from multiple sources, and hand-linking this to form comprehensive family histories, continued for over ten years. Cases of maltreatment were identified and the personal and familial background elaborated by details from relevant files and from the people involved. In parallel with this, departments and agencies were approached to scan their records for new cases or for more information on known or suspected cases. The time spent on completing the search for data for just one child was often counted in days or weeks, which, together with the delays in obtaining records, often hindered the rapid clinical action needed in certain situations. Moreover, considerable clerical effort was involved in separating the material on episodes of maltreatment from dozens, hundreds, or sometimes thousands of documents devoted to other matters in a child's personal or family files.

Although 'unit' record systems exist throughout the health service, there are few opportunities for all records of given conditions, made at different times and places and by different professional groups, to be brought together according to the individual or family affected (Oliver *et al*. 1974). Attempts to collate these records are often hindered by inadequate means of identification, particularly where the individuals and families frequently change names or addresses. These are problems faced by both manual and computerized methods of record linkage, as indicated earlier in the book (Chapter 2), and means of overcoming them have been proposed in the course of the Oxford Record Linkage Study (Acheson 1967; Baldwin 1972, 1973), and by other studies of person and family record linking (Newcombe 1967).

The main focus of attention in this study was children born and maltreated between 1960 and 1980. Most of the first ten-year period was studied retrospectively, collating detail from old files, while the period 1970 to 1980 was considered prospectively, often with active clinical involvement. However, this division is somewhat artificial in that many of the children seen as 'new cases' during the later period, or indeed the sibs of these children, had also been neglected or otherwise maltreated before 1970. The main factors that determined the selection of families were lack of knowledge about the parents' childhood and a failure to meet the criteria for multi-agency support. The latter requirement ensured that the study group did not include families who experienced problems solely with child-rearing. Moreover, it is probable that we would have missed certain of the families in which there were two generations of child maltreatment if there had not been considerable agency involvement.

As a consequence of these factors, it is probable that the selected families were biased towards social classes IV and V and towards the 'indigenous' population, while under-representing recent residents of the area, particularly immigrants (Oliver and Graham 1985). Furthermore, the study was restricted to cases ascertained by July 1981, and subsequent information has revealed that about thirty additional families could have satisfied the selection criteria but were not included in the results.

Results

By July 1981, 147 families fulfilled the criteria for selection. Certain of these families were interrelated and could be grouped together to form 21 kindreds. The 147 mothers had 138 husbands or common-law partners who had fathered at least one of their children and who had lived with them continuously for more than three months. Nine of the mothers never had a male partner satisfying the above criteria. Overall, the women had 74 further husbands and 151 further cohabitees, making 225 additional male partners (363 in all) during 21 years.

Six hundred and sixteen children had been born to the 147 mothers by July 1981, an average of 4.2 children for each woman. Since then, the number has continued to rise, which is to be expected given that the average age of the mothers at the end of 1980 was only 35 years. A further three children had been adopted by the women in the study group. Only 560 of the ascertained children were born between the reference period 1960 to 1980. By the end of 1980, 93 per cent of these children were still alive, with a median age of 12 years and a sex ratio of 114 boys to every 100 girls. Within the 147 families, births were unspaced—almost half of the 560 children were born either in the same year or in the adjacent year to one of their sibs or half-sibs.

Maltreated children

Of the 560 children, 513 had suffered maltreatment ranging from episodic neglect and/or excessive beatings causing surface injuries, through more serious cruelty, to very serious emotional and physical injury leading, in certain cases, to brain damage or death. Some children were rendered epileptic, subnormal, or severely subnormal in intelligence (violence-induced mental handicap: Oliver 1975; Buchanan and Oliver 1977).

The favoured term 'non-accidental injury' is a restrictive euphemism which fails to encompass a whole range of maltreatment, including psychological cruelty, neglect, and 'failure to thrive' (for non-medical reasons or child starvation). More importantly, 'non-accidental injury' is a term which excludes forms of cruelty or assault that leave no obvious marks. At least 37 of the 513 children had been subjected to suffocatory techniques—being held under bath or lavatory water, throttling, having adult hands or pillows held over crying faces, or even having polyethylene bags used punitively or impulsively by parents. Furthermore, many babies had been surreptitiously and violently shaken, sometimes to the point of brain damage or causing inhalation of vomit.

Dead children

Forty-one children out of the 560 born between 1960 and 1980 died within this period. An additional seven child deaths were alluded to in the records but were unconfirmed. Children from the 147 families who died before 1960 or in 1981 were excluded from the analysis. In three cases, the parents were convicted of manslaughter or infanticide. Ten of the deaths were those of index children, two of which led to convictions for infanticide. The child whose death resulted in a verdict of manslaughter was the brother of the two index children for that particular family.

Table 16.1 presents details of the age and sex of the dead children, and gives the death rates per 1000 children at each age. All 41 children died before their eighth birthday. The cumulative mortality rate for the first eight years of life was 77.9 per 1000 children. Nearly two-thirds of the children died between their fifth week and first birthday, the time interval generally recognized as that in which 'baby battering' is most prevalent. The post-neonatal death rate (29–365 days) was an alarming 47.6 per 1000 children. In order to produce a maximum estimate of the 'expected' number of post-neonatal deaths, all mothers in the study group were assumed to belong to social class V, defined according to the Registrar-General's classification of occupations (Office of Population Censuses and Surveys 1970). Data from the Oxford Record Linkage Study for Oxfordshire were used to provide a standard rate. There were 26 observed postneonatal deaths compared with six expected. In other words, the risk of death in the study group was over four times that expected for children born to mothers in social class V (compare Roberts, Lynch, and Golding 1980).

Causes of death

Table 16.2 summarizes the direct causes of death, as stated on the death certificates, for the 41 children. Detailed, collated, confidential information

Table 16.1   Age at death by sex, and death rates per 1000 children at each age

Age at death	Number of deaths				Death rates per 1000 children at each age[1]	Cumulative death rates per 1000 children at each age
	Boys	Girls	Sex unknown	Total		
0–28 days	3	2	1	6	10.9	10.9
29–365 days	14	12	0	26	47.6	58.5
1 year	1	0	0	1	2.0	60.5
2 years	2	1	0	3	6.0	66.5
3 years	0	1	0	1	2.1	68.6
4 years	0	0	0	0	—	68.6
5 years	1	0	0	1	2.2	70.8
6 years	0	2	0	2	4.6	75.4
7 years	0	1	0	1	2.5	77.9
Total	21	19	1	41	—	77.9

[1]   Derived from the number of deaths at each age, occurring between 1960 and 1980, divided by the number of children in the study group surviving and followed-up to each age.

suggests that parental behaviour towards some of these children, and particularly some of those who died in the postneonatal period, had caused or contributed to their deaths. Table 16.3 cross-tabulates the coroners' verdicts of the causes of death with research findings relevant to the circumstances of death. In at least 17 of the 30 deaths officially attributed to misadventure, accidents, or natural causes, there is some evidence that the child was maltreated prior to or at the time of death. Furthermore, it is possible to aggregate the 41 deaths according to a gradation between the three legally-confirmed killings and purely 'biological' causes:

(1) Deaths leading to convictions for manslaughter or infanticide (three cases).

(2) Blatant killings that were unrecognized by the law but were known to certain doctors, to other professionals, and/or to relatives of the dead child (four cases in total: three with surface or hidden injuries to the head and/or neck, including two cases with multiple old and new fractures, and one case of severe neglect with violence).

(3) Instances where the child was killed by a parent actively or by neglect (six cases). Here the official cause of death could only be attributed to accident or illness, but evidence from professionals indicated a killing, usually also with one or more of the following: parent admitted suffocating, or a previous suffocation, or both; relative or cohabitee described the killing; child starved, battered or maltreated in the days

Table 16.2 Age and cause of child deaths

Age at death	Direct cause of death[1]						All causes
	Asphyxia or inhalation of gastric contents	Chest or upper respiratory tract infections[2]	Direct brain damage	Multiple burns or scalds	Other causes[3]	Information not available to research	
0–28 days	—	2	—	—	1	3	6
29–365 days	11	5	3	1	5	1	26
1 year	—	1	—	—	—	—	1
2 years	1	—	—	1	1	—	3
3 years	—	1	—	—	—	—	1
4 years	—	—	—	—	—	—	—
5 years	—	—	—	1	—	—	1
6 years	1	—	—	1	—	—	2
7 years	—	1	—	—	—	—	1
All ages	13	10	3	4	7	4	41

[1] Stated on the death certificate as the disease or condition directly leading to death.

[2] In six cases of chest infection, the underlying causes of death were as follows: *past* (i.e. more than one year previous) brain damage (3 cases); *recent* brain damage (1 case); malnutrition (one case); multiple congenital defects (one case).

[3] Including gastroenteritis (one case), hypothermia (one case), microcephaly (one case), congenital heart disease (one case), neuroblastoma (one case), and 'cot deaths' (two cases).

Table 16.3 Coroners' verdicts of the causes of death compared with research findings relevant to the circumstances of death

Research findings relevant to the circumstances of death	Coroner's verdict of the cause of death			No coroner's inquest or no record or an inquest[1]	Information not available to research	Total
	Infanticide or manslaughter	Open verdict	Misadventure or accidental death			
'Classically battered' babies[2]	1	2	—	—	—	3
Deaths through obvious maltreatment	1[3]	—	1	—	4	
Maltreatment or neglect causing or contibuting to deaths, as supported by collated research data	—	1[3]	6	2	1	10

						Total
Violence-induced mental handicap years previous to death	—	—	—	2	—	2
Bruising and neglect at autopsy	—	—	1[4]	—	—	1
Previous maltreatment, unrelated to death	—	—	3	2	—	5
No maltreatment known	—	—	6	7	1	14
Information not available to research	—	—	—	—	2	2
Total	3	4	16	14	4	41

[1] Deaths certified as natural causes or accidents.

[2] Cases where there were multiple old and new fractures, and surface or hidden injuries involving the head and/or neck.

[3] Case subsequently tried and the accused convicted of cruelty at the time of death.

[4] Three further children showed bruising and evidence of neglect at autopsy, but these are included with the six cases above because collated research data provide evidence for death through maltreatment.

or weeks before death; history of violence-induced mental handicap, the chronic brain damage rendering the child susceptible to pneumonia (Buchanan and Oliver 1977). Past or recent, the maltreatment was related to the death.

(4) Maltreatment or blatant neglect permitted, hastened, or indirectly initiated the natural causes of death (four cases).

(5) Tragic accidents or illnesses, but in young children poorly cared for or left unattended (12 cases).

(6) Prematurity, accidents, or illnesses which may have occurred with no parental responsibility for the death, even if there was maltreatment of the child or his/her sibs (six cases). The remaining six study cases could not be classified precisely but probably belong in one of the last three categories.

Mentally handicapped children

Of the 519 children born to the selected families between 1960 and 1980 and alive at the end of this period, at least 24 were known to be severely subnormal (IQ 0–49), 53 subnormal (IQ 50–75), and 68 specified as dull or borderline subnormal. This classification was based on the most recent available IQ tests and assessments. Table 16.4 cross-tabulates degree of subnormality with data on certain types of maltreatment and physical conditions amongst the children.

By the end of 1980, the point prevalence of severe subnormality among surviving children was 46.2 per 1000. For those children born between 1960 and 1968 and surviving to their 12th birthday, the prevalence of severe subnormality was 60.4 per 1000. This may be contrasted with a prevalence of severe subnormality in the general population of eleven-year-old children of 3.3 per 1000 (Peckham and Pearson 1976; Birch *et al.* 1970; Office of Health Economics 1973). Table 16.4 also indicates that nearly 17 per cent of the surviving children in the severely subnormal intelligence range were known to have suffered violence-induced mental handicap, while a further 75 per cent were reported to have been physically maltreated or neglected.

Fifty-three of the 519 surviving children were recorded as intellectually subnormal (IQ 50–75). This gives a point prevalence of subnormality for the end of 1980 of 102.1 per 1000 children. Among the 265 children surviving to the age of 12 years, 37 were subnormal (139.6 per 1000 children). These findings may be compared with a prevalence of subnormality among eleven-year-olds generally of 17.4 per 1000 (Peckham and Pearson 1976). Compared to the severely subnormal, a lower proportion (nine per cent) of the subnormal children were known to have violence-induced mental

Table 16.4 Mental handicap in relation to certain types of maltreatment and physical conditions among surviving[1] children

	All cases of mental handicap[3]	Number of cases[2]					
		Violence-induced mental handicap[4]	Deprivation and head injuries	Deprivation reported but no head injuries	No Deprivation or violence to the head	'Failure to thrive' specified	Recurrent Epilepsy
Severe Subnormality (IQ 0–49)	24 (4.6%)	4	10	8	2	5	7
Subnormality (IQ 50–75)	53 (10.2%)	5	29	15	4	12	10
All mentally handicapped	77 (14.8%)	9	39	23	6	17	17

[1] Children born between 1960 and 1980 and alive at the end of this period. A further six severely subnormal children and one subnormal child from the study group died within the reference period.
[2] First four categories are mutually exclusive.
[3] Bracketed figures are mentally handicapped cases as a percentage of all surviving children.
[4] Minimum number of VIMH (i.e. medically-confirmed cases only).

handicap, but a slightly greater proportion (83 per cent) were reported to have been physically maltreated or neglected.

Causes and contributory factors

The surviving cases of severe subnormality may be classified into recognized aetiological categories (Crawfurd 1982) and cross-referenced with the probable causes or contributory factors as indicated by the research data (Table 16.5). It can be seen that ten of the cases have been assigned, albeit cautiously, to the category 'predominantly environmental causes'. Three of the children clearly suffered violence-induced mental handicap as a result of battering, violent shaking, and/or other physical abuse when they were just a

Table 16.5 Classification of causes of severe subnormality (IQ 0–49) among surviving[1] children

Standard aetiological categories for mental handicap	Cause supported by research data—Number of cases[2]			
	Violence-induced mental handicap[3]	Maltreatment/ deprivation a major contributant to the mental handicap	No strong evidence that neglect or violence was a major contributant to the mental handicap	Total
Predominantly environmental causes	4	4	2	10
Chromosomal defects	0	0	2	2
Single gene defects	0	0	0	0
Multifactorial or of unknown aetiology:				
(a) with stigmata or other findings of likely congenital origin	0	0	4	4
(b) with psychotic features	0	2	2	4
(c) no stigma or psychotic features	0	2	2	4
Total	4	8	12	24

[1] Children born between 1960 and 1980, and alive at the end of this period.
[2] Mutually exclusive categories.
[3] Minimum number of cases of VIMH (i.e. medically-confirmed cases only).

few months old (Oliver 1975). To these three should be added a fourth child, who developed meningitis and brain damage as a consequence of a fractured skull following parental violence and neglect (see, for comparison, Akuffo and Sylvester 1983; Eppler and Brown 1977). A further child was given a diagnosis of 'probable encephalitis' but had also been repeatedly and very severely neglected. There were three children in which the severe mental handicap had been assigned by paediatricians and a child psychiatrist to starvation or severe social deprivation, or both. The developmental quotients of these children were very low but rose when the children were placed in safer, happier surroundings. However, the most recent IQ measurements available to the study indicated that these children were still within the range of severe subnormality. Finally, there were two cases of spastic quadriplegia attributed to 'prematurity and asphyxia' and 'neonatal anoxia', but these diagnoses were retrospective and uncertain.

There are particular problems with attempting a similar classification for the subnormal children. The causes here are mostly uncertain and often multiple, with probable synergistic interactions between inherited dispositions, adverse environmental influences due to disease or detrimental social pressures, as well as obvious and less obvious forms of violence-induced mental handicap. Of the 53 children surviving by the end of 1980 and regarded as subnormal, there were five cases of violence-induced mental handicap (compare Developmental Medicine and Child Neurology 1983; Diamond and Jaudes 1983). Four of the five children had suffered brain damage at the hands of a parent on one or more occasions, with obvious fulminant episodes. The other child had received extreme and prolonged violence to the head for about three years, but without proven subdural or subarachnoid haemorrhages (Buchanan and Oliver 1977). Further details on the maltreatment of these five children and on the remaing 48 cases with subnormal intelligence are given in Oliver and Graham (1985).

Parental violence and child epilepsy

The research data not only suggest a link between parental violence and mental handicap in children, but also between parental violence and a tendency for the children to have epileptic fits (see also Caffey 1972; Lancet 1974; Cooper 1978). Table 16.4 indicated that 22 per cent of the 77 mentally handicapped children alive at the end of 1980 were, at some point before the age of 12 years, reported to have recurrent epilepsy; to these may be added a further 14 intellectually normal children with epilepsy. Moreover, four of the children who died within the study period were diagnosed as epileptic during their lifetime. One of these deaths was in fact attributed to status epilepticus. Overall, 35 of all 560 children were diagnosed, before the age of 12 years, as having recurrent epilepsy, which gives a prevalence of 62.5 per 1000 children (compare Ross *et al*. 1980; Shorvon 1984).

Discussion

This study focused on 147 families in north-east Wiltshire in which at least two generations have suffered from neglect or abuse during childhood. For 13 years, the author (JEO) was immediately concerned with at least one member of most of the families in the study and, as a psychiatrist, was told personal accounts of killings or violence-induced mental handicap suffered by the children. In addition to this 'inside' information, supporting data was collated on the families, including unequivocal direct statements as well as the more common partial or attenuated admissions. Detail came from the mothers themselves, other relatives, or professional people involved in child welfare.

Dead and discarded children tend to be omitted from the current medical and social records of problem families. Moreover, certain children who are fostered, or in care, or in institutions, may be forgotten, especially if they are mentally handicapped. This study found that information about dead or damaged children was also often withheld or forgotten by the mothers themselves, particularly if the child was only the half-sibling of the children remaining in the family. If facts as basic as the life or death of the sibs or half-sibs of a maltreated child are unknown to child welfare agencies, then it is also probable that many cases of child maltreatment pass unnoticed. More effective family-linked information could help ensure better protection and happier lives for the children at risk.

The need for family maltreatment registers

The resources required for linking information in this particular study were considerable and the methods used could not easily be replicated on a routine basis in other parts of the country. However, the findings do indicate a need for family-based case registers of child maltreatment. These registers could consist of abstracts of events sorted on the basis of the maltreated child's mother's birth surname and cross-referenced with father's surname. Each abstract would include the names of the child's sibs and half-sibs at the time of the event. The records for related children who had been maltreated would be linked together to produce family histories. Information need not be confined to events or episodes of maltreatment. The study in north-east Wiltshire has demonstrated the importance of information from agencies that provide families with social support. Each contact with an agency made by a family on the child maltreatment register would generate a record. These various items of information about the children and their families, when linked together, would assume far greater significance than when considered separately; this is the basic idea underlying record linkage (Acheson 1967).

The benefits to be gained from this approach not only include the ascertainment of abused children, their treatment, and the secondary prevention of further damage. The linked data also emphasize a special feature of this problem, namely that abuse is but a manifestation of widespread, heterogeneous, and often severe, medical and social pathology affecting the family as a whole. As such, child abuse is unlikely to be controllable, much less preventable, if treated as an isolated event.

Acknowledgements

Appreciation is given to co-workers who were associated with earlier research contributing to this chapter. Particular thanks are due to Mrs Audrey Taylor, Mrs Jane Cox, Mrs Anne Buchanan, and to the late Dr John Baldwin.

Parts of this chapter are reprinted from the *British Medical Journal* (1983), **286**, 115–17, and from Oliver and Graham (1985).

References

ACHESON, E. D. (1967). *Medical record linkage*. Nuffield Provincial Hospital Trust. Oxford University Press.

AKUFFO, E. O. AND SYLVESTER, P. E. (1983). Head injury and mental handicap. *Journal of the Royal Society of Medicine* **76**, 545–9.

BALDWIN, J. A. (ed.) (1972). *The master index: a guidebook for medical record officers*. Oxford Regional Hospital Board.

—— (1973). Linked record medical information systems. *Proceedings of the Royal Society of London* **184**, 403–20.

BIRCH, H. G., RICHARDSON, S. A., BAIRD, D., HOROBIN, G., AND ILLSLEY, R. (1970). *Mental subnormality in the community*. The Williams and Wilkins Co., Baltimore.

British Paediatric Association (1966). 'The battered baby.' Memorandum by the special standing committee on accidents in childhood. *British Medical Journal* **1**, 601–3.

BUCHANAN, A. H. AND OLIVER, J. E. (1977). Abuse and neglect as a cause of mental retardation. *British Journal of Psychiatry* **131**, 458–67.

CAFFEY, J. (1946). Multiple fractures in the long bones of infants suffering from chronic subdural haematoma. *American Journal of Roentgenology* **56**, 163–73.

—— (1972). On the theory and practice of shaking infants; its potential residual effects of permanent brain damage and mental retardation. *American Journal of Diseases of Children* **124**, 2, 161–9.

CHRISTOFFEL, K. K., LIU, K., AND STAMLER, J. (1981). Epidemiology of fatal child abuse: international mortality data. *Journal of Chronic Diseases* **34**, 57–64.

COOPER, C. E. (1978). Child abuse and neglect—medical aspects. In *The maltreatment of children* (ed. S. M. Smith) pp. 9–68. MTP Press Ltd., Lancaster.

CRAWFURD, M. D'A. (1982). Severe mental handicap: pathogenesis, treatment, and prevention. *British Medical Journal* **285**, 762–6.

Developmental Medicine and Child Neurology (1983). Child abuse and cerebral palsy (editorial) **25**, 141–2.

Diamond, L. J. and Jaudes, P. K. (1983). Child abuse in a cerebral palsied population. *Developmental Medicine and Child Neurology* 25, 169–74.

Eppler, M. and Brown, G. (1977). Child abuse and neglect. Preventable causes of mental retardation. *First International Conference on Child Abuse and Neglect* Geneva.

Griffiths, D. L. and Moynihan, F. J. (1963). Multiple epiphysial injuries in babies ('battered baby' syndrome). *British Medical Journal* 2, 1558–61.

Lancet (1974). Children in danger (editorial) 1, 1090–1.

Newcombe, H. B. (1967). Record linking: the design of efficient systems for linking records into individual and family histories. *American Journal of Human Genetics* 19, 335–59.

Office of Health Economics (1973). *Mental Handicap*. OHE booklet No. 47. White Crescent Press Ltd., Luton.

Office of Population Censuses and Surveys (1970). *Classification of Occupations*. HMSO, London.

Oliver, J. E. (1975). Microcephaly following baby battering and shaking. *British Medical Journal* 2, 262–4.

—— and Cox, J. M. (1973). A family kindred with ill-used children: the burden on the community. *British Journal of Psychiatry* 123, 81–90.

—— and Graham, W. J. (1985). Generations of maltreated children in north-east Wiltshire. *Oxford Record Linkage Study Research Report*.

—— and Taylor, A. (1971). Five generations of ill-treated children in one family pedigree. *British Journal of Psychiatry* 119, 473–80.

——, Cox, J. M., and Buchanan, A. (1978). The extent of child abuse. In *The maltreatment of children* (ed. S. M. Smith) pp. 121–74. MTP Press Ltd., Lancaster.

——, ——, Taylor, A., and Baldwin, J. A. (1974). Severely ill-treated young children in north-east Wiltshire. *Oxford Record Linkage Study Research Report* No. 4.

Peckham, C. S. and Pearson, R. (1976). Prevalence and nature of ascertained handicap in the National Child Development Study (1958 cohort). *Public Health* 90, 111–21.

Roberts, J., Lynch, M. A., and Golding, J. (1980). Postneonatal mortality in children from abusing families. *British Medical Journal* 281, 102–4.

Ross, E. M., Peckham, C. S., West, P. B., and Butler, N. R. (1980). Epilepsy in childhood: findings from the National Child Development Study. *British Medical Journal* 280, 207–10.

Shorvon, S. D. (1984). Epilepsy update. *Hospital Update* 10 (6), 541–6.

Smith, S. M. (ed.) (1978). *The maltreatment of children*. MTP Press Ltd., Lancaster.

—— and Hanson, R. (1974). One hundred and thirty-four battered children: a medical and psychological study. *British Medical Journal* 3, 666–70.

17

Incidence of cancer in relatives of children with retinoblastoma

J. FEDRICK* AND THE LATE J. A. BALDWIN

Children with retinoblastoma have an increased risk of developing osteogenic sarcoma (Aherne 1974; Kitchin and Ellsworth 1974), and members of families containing children with retinoblastoma may be at increased risk of developing cancer at other sites (Aherne 1974; Chan and Pratt 1977; Gordon 1974), though evidence for this has been anecdotal. We have therefore investigated whether families containing an infant with retinoblastoma are prone to cancer.

Methods and results

With permission from the consultants, we examined the files of the Oxford Regional Cancer Registry for all children in the region known to be alive who had been diagnosed as cases of retinoblastoma. The general practitioner in each of these index cases was contacted and permission sought to interview the mother at home. Information obtained on the families included full names of first-, second-, and third-degree relatives, their places of residence since 1952, and their dates of birth. The files of the Regional Cancer Registry and the Oxford Record Linkage Study were then searched for any cases of cancer among the family members who had been resident in the region at any time from 1952. The expected numbers of relatives with cancer were calculated by using the period during which each person was at risk together with published data on the incidence of the disease related to age and sex in the region (Doll, Muir, and Waterhouse 1970).

Eleven families were interviewed. In only one index case did the child have bilateral disease. The number of relatives who had developed cancer was significantly higher than expected (see Table 17.1). Details of the families are given below.

Family 2—Index case was a girl with retinoblastoma of the left eye diagnosed at age 5, and tumour in the other eye diagnosed five months later. She was mentally retarded with odd facies and had congenital dislocation of the hips. Her paternal great-grandfather had died of lung cancer at age 67.

* Now Golding.

Reprinted from the *British Medical Journal* 1 , 83–4 (1978).

Table 17.1 Observed and expected numbers of cases
of cancer in relatives resident in Oxford region
since 1952

Relationship to index case	No. observed	No. expected
Great-grandfather	1	0.116
Great-grandmother	0	0.136
Grandfather	1	0.199
Grandmother	2	0.571
Father	1	0.178
Mother	0	0.283
Brother	0	0.032
Sister	0	0.015
Total	5*	1.530
Relative risk		3.3

* $P = 0.01$ (One-tailed test used; expectation assumed
to be mean of Poisson distribution.)

Family 4—Index case was a girl with retinoblastoma of the left eye diagnosed
at age 3. Her maternal grandmother had developed breast cancer at age 58.
Family 6—Index case was a girl with retinoblastoma of the right eye
diagnosed at age 1. Her paternal grandfather had developed stomach cancer
at age 67, and her paternal grandmother had developed breast cancer at age
68.
Family 10—Index case was a boy with retinoblastoma of the right eye
diagnosed at age 1. His father had died of bronchial carcinoma at age 46.

During the interviews it became apparent that other members of the 11
families had also developed cancers, although this information was not
specifically asked for. In most instances the diagnosis had been made abroad
and so could not be verified. Interestingly, however, a distant paternal cousin
of the index case in family 6 was said to have had an eye removed in childhood, and a maternal first cousin of another index case was said to have died
of cancer of the leg at the age of 13.

Comment

The 11 families interviewed were typical of the Oxford region in that they
were highly mobile. By confining our analysis to information on relatives
resident in the region since 1952 we restricted the data set. Nevertheless, a
significant excess of relatives with carcinoma was found.

Acknowledgements

We thank the consultants and general practitioners for permission to interview the parents, Dr A. Barr for allowing us to use the data of the Oxford Regional Cancer Registry, and Miss C. Hunt for co-operation. Interviews were conducted by Mrs Yvonne Timms and Mrs Bettine Sutton.

References

AHERNE, G. (1974). Retinoblastoma associated with other primary malignant tumours. *Transactions of the Ophthalmological Societies of the UK* **94** , 938.

CHAN, H. AND PRATT, C. B. (1977). A new familial cancer syndrome. A spectrum of malignant and benign tumours including retinoblastoma, carcinoma of the bladder and other genitourinary tumours, thyroid adenoma, and a probable case of multifocal osteosarcoma. *Journal of the National Cancer Institute* **58** , 205.

DOLL, R., MUIR, C., AND WATERHOUSE, J. (1970). *Cancer Incidence in Five Continents* Vol. 2. Springer-Verlag, Berlin.

GORDON, H. (1974). Family studies in retinoblastoma. *Birth Defects Original Article Series* **10** No. 10, 185–90.

KITCHIN, F. D. AND ELLSWORTH, R. M. (1974). Pleiotropic effects of the gene for retinoblastoma. *Journal of Medical Genetics* **11** , 244.

PART III

Implications of record linkage

18

Implications of record linkage for health services management

M. J. GOLDACRE

Introduction

The main function of management in the health service is to provide an appropriate framework within which effective health care can be obtained by those in need. Since resources are limited, it is also management's job to ensure that resources are equitably distributed between competing demands, and to ensure that health care facilities are used efficiently so that the use of services is maximized.

To help meet these ends, management requires information about both the health needs of the community and the use of health services by the people it serves. It also requires information not only to monitor the current performance of its services but also to guide decisions about future developments in health care and to monitor the results of policy decisions.

Uses of numerical information in health service management have been described in a number of publications (Ashley 1972; Benjamin 1971; MacKay and Charles 1951; Nuffield Provincial Hospitals Trust 1977). Such information rarely provides complete answers to complex problems. In general, the role of information is to reduce the uncertainties inherent in decision-making, to confirm or correct impressions gained from subjective observation and experience, and to alert management to unusual or unexpected circumstances. This chapter is concerned with ways in which the use of linked records can improve the quality of information for management compared with data based on unlinked records. The examples are drawn mainly from the Oxford Region and are concerned, in particular, with data which distinguish counts of individual patients from counts of episodes of health care.

Local comparisons with national in-patient admission rates and lengths of stay

Few measures available to health service managers are absolute. The quantification of health needs remains elusive, demands for health care show considerable elasticity, and 'norms' for the provision of beds and manpower are usually based on averages and/or projections of current practice. Managers therefore frequently make use of comparative statistics to improve their understanding of the performance of local health services. Admission to in-patient care and the length of time spent in hospital are the two determinants of the use of hospital beds, and thus it is not surprising that hospitalization rates and lengths of stay often receive close scrutiny. Indeed, one of the main administrative uses of the Hospital In-patient Enquiry is to facilitate regional comparisons of bed usage, discharge rates, and lengths of stay (Department of Health and Social Security 1977).

A striking feature of the provision of hospital services in the Oxford Region is the low availability of hospital beds per unit of population compared with the country as a whole (Table 18.1). It is, therefore, important locally to determine whether fewer patients receive hospital in-patient care in the Oxford Region than elsewhere. Table 18.1 compares hospital discharge rates in Oxford with those for England and Wales. (The population of Oxford is a little younger than that nationally: the standardized discharge ratio compares the observed number of discharges in the region with the number which would have been expected if the local population had experienced the age- and sex-specific discharge rates of the country as a whole.) The table shows that, although the per capita number of beds for general specialties is about 18 per cent lower than the national average, the overall discharge rate was similar to the national average, both in absolute terms and relative to the age-sex structure of the population.

It is obvious that average or higher-than-average discharge rates, with a lower than average provision of beds, can only be achieved by shorter lengths

Table 18.1 Availability of beds and discharge rates per 1000 population for selected specialties in the Oxford Region and in England and Wales, 1977.

Specialty	Beds (per 1000)		Discharges (per 1000)		S.D.R.*
	Oxford	England and Wales	Oxford	England and Wales	
All general specialties†	3.80	4.62	88.7	89.7	104.3
General medicine	0.44	0.63	14.3	16.2	93.5
Geriatrics	0.90	1.21	4.9	4.9	128.7
General surgery	0.47	0.62	18.3	19.9	98.0
ENT surgery	0.09	0.11	5.6	4.8	121.8
Trauma and orthopaedic surgery	0.43	0.45	11.4	9.1	140.7
Ophthalmology	0.06	0.09	2.0	2.5	89.3
Gynaecology	0.18	0.22	11.0	10.3	102.5

* Standardized discharge ratio for Oxford (England and Wales = 100): ratio of the observed to expected number of discharges in Oxford, standardized to age- and sex-specific rates.

† Excludes maternity and psychiatry.

of stay or higher bed occupancy rates than those nationally. In fact, lengths of stay are generally shorter in Oxford than elsewhere, both absolutely and when standardized for differences in age structure between the local and national hospitalized population (Table 18.2).

Table 18.2 Average lengths of stay in selected specialties in Oxford and in England and Wales, 1977

Specialty	Average length of stay (days)		S.L.O.S.R.*
	England and Wales	Oxford	
General surgery	8.5	6.4	92.1
Trauma and orthopaedic surgery	14.3	10.0	74.1
ENT surgery	4.2	3.1	73.8
Ophthalmology	7.4	4.7	63.5

* Standardized length of stay ratio (England and Wales = 100) : ratio of the actual to the expected length of stay in Oxford, standardized to national figures for each age-sex group.

Shorter stay—real or apparent?

A number of factors must be taken into account when considering discharge rates or lengths of stay (Goldacre 1981). Among these is the fact that figures which are available nationally relate to episodes of hospital care rather than to individual people. Thus an individual patient who is transferred between hospitals or readmitted to hospital is counted twice (or more) in the statistics. The discharge rates in Table 18.1 do not refer to 'people treated' but to 'episodes' of hospitalization, including transfers and readmissions. So can we infer that the number of people treated in general surgery in Oxford is similar to the number treated nationally, despite Oxford's 25 per cent deficit of beds? Similarly, can we infer that, although the bed complement available to geriatricians and orthopaedic surgeons is lower than the national average, the number of individuals treated by them is considerably higher?

Alternatively, is it possible that (perhaps *because of* the relatively low local bed-provision in the region) pressures to transfer patients between hospitals or to discharge them early and then readmit them are greater in Oxford than elsewhere? The comparatively shorter lengths of stay in Oxford than the national average are, of course, compatible with either explanation. The distinction between these two possibilities—more individuals treated or simply more transfers and readmissions—is crucial to understanding any comparison between local and national figures. At the extremes of interpretation, the figures for Oxford may represent a service working efficiently and requiring fewer beds than the national average, or it may represent a service working under strain. In the absence of nationally available, person-based data this is a distinction which cannot be made from hospital statistics.

Readmission rates and transfers: a local example

The potential influence on comparative statistics of readmissions and transfers, described above, is far from theoretical because the scale of readmissions and transfers can be considerable. Table 18.3 shows, from linked records within the Oxford Region (Baldwin, Golding, and Simmons 1980), the probability of readmission to hospital at fairly short time-intervals after discharge. Table 18.4 shows the number of people 'discharged' from hospital by being transferred from one hospital to another. The problem of

Table 18.3 Probability of readmission after discharge from selected specialties: cumulative percentage of patients readmitted within one week and within three months from date of discharge

Specialty	Period after discharge	Males (%)		Females (%)	
		Oxfordshire	West Berkshire	Oxfordshire	West Berkshire
General surgery:	within 1 week	8.8	16.1	7.5	13.0
	within 3 months	16.0	22.0	14.5	19.3
Trauma and orthopaedics:	within 1 week	5.0	12.6	7.6	17.0
	within 3 months	11.9	17.5	15.6	23.0
Gynaecology:	within 1 week	—	—	16.7	6.9
	within 3 months	—	—	22.2	12.2

readmission rates and transfers is aggravated because, as Tables 18.3 and 18.4 show, they vary from place to place. In addition, transfers are subject to an important anomaly in unlinked statistics at present. If a patient is transferred between two specialties (say, orthopaedic surgery and geriatrics) within the same hospital, the patient's stay is counted as one hospital episode. If, however, the specialties are on two different sites requiring a transfer between two hospitals, two episodes are counted in the statistics thus inflating the discharge rate and reducing the average length of stay.

Lengths of stay and throughput per bed

Lengths of stay and throughput per bed (the number of patients treated per available bed per unit of time) are two measures of hospital 'efficiency' which are much used in health service management. The following example

Table 18.4 Hospital discharges, transfers to other hospitals, and transfers as a percentage of all hospital discharges

Speciality	Northampton			Wycombe		
	Discharges	Transfers	T/D (%)	Discharges	Transfers	T/D (%)
General medicine	4458	673	15.1	3512	171	4.9
General surgery	6489	109	1.7	5293	71	1.3
Trauma and orthopaedics	5204	544	10.5	2403	84	3.5
Paediatrics	2950	769	26.1	2087	56	2.7
Gynaecology	3704	19	0.5	2648	22	0.8
Geriatrics	691	100	14.5	1643	107	6.5

emphasizes the caution that is needed in interpreting such statistics based on unlinked records. The two hospitals in Table 18.5 are both geriatric hospitals within the Oxford Region. It is clear from the episode-based statistics that there are substantial differences between the hospitals in the management of their patients although it is not clear how or why they differ. The average length of stay in hospital A was 29 days compared with 120 days in hospital B; the number of hospital episodes per bed was respectively 12 and 3 per year.

Table 18.5 Lengths of stay and throughput per bed based on episodic data

	Hospital A	Hospital B
Available beds	193.8	106.7
Discharges-and-deaths during year	2302	312
Average length of stay (days)	29.3	120.3
Throughput per bed	11.9	2.9

A much fuller picture of the difference between the hospitals emerges from statistics based on counts of individuals rather than episodes (Table 18.6). The average number of admissions per person during the year was 4.3 in hospital A (with some patients being discharged and readmitted at fortnightly or weekly intervals) compared with 1.3 in hospital B. The average time spent in hospital per individual patient during the year was 126 and 156 days respectively, and the number of individuals treated per hospital bed during the year was 2.8 and 2.3. It is clear that the policy in hospital A is, for at least a proportion of their patients, to discharge and readmit them. Interpretation of administrative statistics based on episodes alone is likely to become more difficult as the implementation takes place of such policies as planned discharge and readmission, five-day wards, transfers between specialties to share clinical care, and transfers between specialties and hospitals to release high-dependency beds.

Table 18.6 Lengths of stay and throughput per bed based
on data about individuals

	Hospital A	Hospital B
Available beds	193.8	106.7
People treated during year	539	241
Average time in hospital per person (days)	125.6	155.7
People treated per bed	2.8	2.3

Trends in admission rates: national figures

The number of hospital discharges-and-deaths per 1000 population in England rose by 30 per cent between 1959 and 1977 (Table 18.7) with a corresponding fall of about one-third in the average length of time spent in hospital per episode. The problems of interpreting trends in these figures over time are similar to those, described above, of interpreting variation in discharge rates and lengths of stay in regional or sub-regional comparisons. The first problem is whether the trends mainly reflect an increase in the number of individuals treated in hospital or simply an increase in re-admissions and transfers. Data from the Mental Health Enquiry (Office of Population Censuses and Surveys 1977), which distinguish between first admissions and readmissions, indicate that the increase in admissions to psychiatric hospitals in recent years reflects an increase in readmission rates and that the rate of first admissions to psychiatric care has, in fact, declined. These findings cannot necessarily be extrapolated to other specialties.

Table 18.7 Discharges and deaths
per 1000 population in England,
1959–77 (all specialities)

Year	Rate (per 1000)
1959	88.5
1963	97.3
1967	104.0
1970	108.4
1971	112.2
1972	112.9
1973	110.5
1974	111.4
1975	107.1
1976	113.2
1977	115.3

The distinction between a rise in the number of readmissions and transfers and a rise in the number of individuals admitted to hospital—whether because of a lowered clinical threshold for admission or an increase in the number of individuals in need of hospital care—is important. It is central to any understanding of how the health service is working and to how its work patterns are changing, to the planning of health care, and to the assessment of the number of individuals in the community in need of hospital care.

Local planning for clinical groups

A common management problem concerns the estimation of the number of people in the population needing specialized services and, particularly, the number of people with chronic conditions, such as multiple sclerosis, haemophilia, cystic fibrosis, or congenital hypothyroidism (Table 18.8). Since the introduction of Hospital Activity Analysis it is easy to count the number of episodes of care, though without record linkage it is not possible to infer the number of individuals in need of care, as measured by hospitalization. For example, the fact that there were 1184 discharges for haemophilia in the Oxford Region between 1974 and 1978 gives no reliable indication of the number of individuals with haemophilia in the population who required hospital care. Linking records showed that the 1184 episodes related to 316 individuals. The episode-based admission rate for haemophilia among male residents of Oxfordshire under 15 years of age was higher than that among residents of the rest of the health region. Oxfordshire contains the regional centre for the treatment of haemophilia and, in the absence of any other data, it might be tempting to guess that this could reflect a higher readmission rate among local residents than among others because of their proximity to the service. In fact, this might not be the explanation. Linking records showed that the person-based admission rate was also higher in Oxfordshire than in the rest of the region. This may represent a higher concentration of children with haemophilia—and a greater need for services—among the resident population of the district than elsewhere (perhaps through immigration of families to take up residence close to the centre).

Table 18.8 Number of hospital episodes and number of people treated in hospital in the Oxford Region, 1974–8

Disease	Episodes	People
Multiple sclerosis	1268	759
Haemophilia	1184	316
Cystic fibrosis	182	96
Congenital hypothyroidism	9	9

Table 18.9 shows that hospital admissions for leukaemia rose sharply in the Oxford Region in recent years. The distinction between a change in the occurrence of the disease and simply a change in readmission practice is of particular interest to the epidemiologist; but the health service manager will also want to know whether there is a need to provide services for an increased number of newly diagnosed cases or for multiple readmissions as the resource implications will differ accordingly. The reason for the rise, as the table shows, was an increase in readmission rates.

Table 18.9 Number of hospital episodes and number of individual patients admitted to hospital with lymphatic and myeloid leukaemia in the Oxford Region, 1974–8

Diagnosis		1974	1975	1976	1977	1978
Lymphatic leukaemia:	episodes	94	170	188	292	280
	patients	62	50	71	65	60
Myeloid leukaemia:	episodes	98	163	198	356	282
	patients	57	77	71	58	71

Strategic planning of hospital beds

One of the major strategic tasks of health service management is the planning and provision of hospital beds. The number of beds required per 1000 population can be expressed as follows:

$$\text{Beds per 1000 population} = \frac{\text{D.R.} \times \text{L.O.S.}}{P \times 365}$$

where D.R. is the estimated discharge rate per 1000 people, L.O.S. is the estimated length of stay, and P is the proportion of beds estimated to be occupied on average per day.

At present, a bed occupancy of about 75 per cent seems generally achievable. The finite time between vacating and filling beds and the inevitable peaks and troughs in admissions mean that one cannot plan for continuous 100 per cent occupancy. The national discharge rate and average length of stay for general specialties are about 88 per 1000 and 13.5 days, respectively. Thus the number of general beds needed per 1000 population on current figures is about $(88 \times 13.5)/(0.75 \times 365) = 4.3$. However, the lag time in any major capital development is usually considerable and strategic planning is concerned with figures which are likely to be relevant at future periods in time. Statistics on the use of hospital beds in the past have shown some striking trends—notably the rising discharge rate and falling lengths of stay—and these need to be taken into account in predicting future bed requirements. A common starting-point is to project past trends in age-, sex-, and specialty-specific discharge rates and lengths of stay, although simple extrapolation is insufficient and must be modified by judgment. It is, for example, self-evident that the falling lengths of hospital stay cannot continue indefinitely.

Judgment about future needs is difficult but necessary and judgment on trends based on unlinked data is particularly difficult. Statistics on discharge rates and lengths of stay include first admissions, relapses, planned re-admissions, and transfers. It is difficult for a clinician or manager to conceptualize, say, what an episode-based discharge rate of 88 per 1000 or a length of stay of 13.5 days really represents. It is difficult to form judgments with any degree of confidence about how they might change, either absolutely or relative to one another. It is a little easier to judge whether the number of people in need of care (say, the number of diabetics or schizophrenics) is likely to change, and whether there are prospects for further reducing individuals' length of stay in hospital. Such judgments would be better made and monitored with the aid of person-based data rather than the complex mix that constitutes episode-based data.

The strategic planner will often apply age-, sex-, and specialty-specific figures from a large population—the region or the country as a whole—to local populations to derive the estimated number of beds required locally. The resultant figures, 'norms' or averages, do not take account of any special local circumstances and thus become the first bargaining point with local management. The pleading of local circumstances, the 'special case' which is so familiar in the health service, is often difficult to confirm or refute from local figures based on the complex case-mix of episode-based data rather than person-based data.

Waiting lists

A seemingly intractable problem in the National Health Service is the size of surgical waiting lists. A recent report on trauma and orthopaedic surgery (Department of Health and Social Security 1981) has drawn attention to the possibility that long lengths of stay for patients with accidents may block beds for cold orthopaedic surgery. It is therefore important to question whether there is scope for reducing the length of stay for patients admitted to hospital for trauma. Table 18.10 compares lengths of stay for fractures of the femur in three hospital regions and the figures suggest about a two-fold variation in length of stay. But at least three interpretations of the figures are possible and the distinction between them, though important, is impossible to make from episode-based data. Firstly, patients may be discharged home earlier in Oxford than in the other two regions. Secondly, patients may be transferred out of orthopaedic beds to other hospitals, thereby releasing orthopaedic beds, more often in Oxford than in the other two regions. Thirdly, patients

Table 18.10 Mean and median lengths of stay in three regions for fractures of the femur

Region	Length of stay (days)			
	Fractured neck of femur		Other fractures of femur	
	Mean	Median	Mean	Median
Oxford	23.0	17.5	43.6	28.0
North Western	42.2	27.6	55.5	42.5
Mersey	49.5	32.3	57.9	46.2

may be transferred out of orthopaedic beds similarly in the three regions, but this may involve transfers between hospitals in Oxford and between specialties within the same hospital in the other two regions.

Geographical resource allocation: patients who cross administrative boundaries

In the United Kingdom, health authorities are reimbursed for the treatment of hospital in-patients whose place of residence is outside their administrative boundaries (Department of Health and Social Security 1976). Reimbursement is made on the basis of episodes treated rather than patients treated. We have seen that readmission policies can vary substantially from place to place (Tables 18.5 and 18.6). If hospital B cares for a non-resident patient in one continuous episode, its district will be reimbursed for the treatment of one episode; if hospital A cares for a similar non-resident patient by, say, four brief episodes of admission, discharge, and readmission, it will be reimbursed fourfold.

Outcome of care

Health service managers need to consider not only the type and scale of services provided but also their effectiveness. The efficacy of a service (say, coronary care units or surgery for certain types of cancer) is concerned with measuring the effect of the service in altering the natural history of the disease for the better under optimum, well-supervised circumstances. It is best measured by randomized controlled trials that address the question: can the service confer benefit? The quality of a service is concerned with whether medical care is applied better in some locations than in others and its measurement addresses the question: does the local service confer greater or less benefit than that available elsewhere? Inferences about quality can, in general, only be made by careful analysis and interpretation of descriptive data used to monitor the performance of services in different locations. The use of unlinked data to monitor the outcome of care is often unsatisfactory because of difficulties of interpretation.

Acheson (1968), for example, noted striking differences in the proportion of patients with fractured neck of femur who died in one hospital compared with another (Table 18.11). In hospital A, 11 per cent of the patients died in the initial admission while in hospital B, 35 per cent died. Linkage of subsequent episodes of in-patient treatment and death certificates to the record of the first admission was therefore undertaken and it was found that there was no overall difference in the mortality experience of the groups treated in the two hospitals.

Table 18.11   Deaths from fractured neck of femur in two hospitals

	Hospital A		Hospital B	
	Number	%	Number	%
Died during first admission	16	11.1	34	34.7
Died during subsequent admission	30	20.8	5	5.1
Died subsequently at home	9	6.3	2	2.0
Subtotal: all deaths	55	38.2	41	41.8
Total: all patients	144	100.0	98	100.0

Other studies (Ashley, Howlett, and Morris 1971; Goldacre 1976; Lipworth, Lee, and Morris 1963) have pointed to differences in fatality rates between teaching and non-teaching hospitals for a number of diseases but, without record linkage or special studies, such differences are difficult to monitor routinely with confidence. For a given condition, the 'hospital fatality ratio' available from unlinked data is the number of hospital episodes which end in death expressed as a ratio of all hospital episodes for the condition. It does not distinguish first episodes from relapses and re-admissions, and it takes no account of deaths which occur in subsequent admissions or outside hospital. When present, however, differences in outcome between hospitals are potentially important in making decisions about the deployment of resources. Morris (1969) has commented that 'one hope of making an impression on death rates is by more and super technology . . . but we do not know how many deaths might have been prevented using available methods'.

Access to care

Linkage of records provides the capacity to follow the care received by individual patients during different hospital episodes for the same spell of illness. It facilitates measurement of the extent of transfers between hospitals for the sharing of care for particular conditions, and it facilitates monitoring of access to specialized care among patients who are initially admitted to general units.

For example, the only neurological and neurosurgical facilities in the Oxford Region are situated in the teaching hospitals in the city of Oxford. We were interested in the extent to which young people with stroke—those under 55 years of age—were transferred for neurological investigation to identify conditions, such as aneurismal subarachnoid haemorrhage, amenable to surgery. (Unlinked data would simply have shown, for example, that no patients admitted to non-teaching hospitals underwent surgery.) In all, taking account of transfers, we found that 48 out of 178 patients (27.0 per cent) initially admitted to non-teaching hospitals and 33 out of 131 patients (25.2 per cent) initially admitted to teaching hospitals were found to have an aneurism on investigation (as confirmed by case-note review). Eighteen (13.7 per cent) of those initially admitted to non-teaching hospitals and 23 (12.9 per cent) of those first admitted to teaching hospitals underwent surgery. It therefore seems reasonable to conclude that, although the neuro-surgical service is centralized in one city, there was reasonable equality of access to it.

Record linkage and the roles of management

To generalize, the roles of management are to assess the adequacy, effectiveness, efficiency, and accessibility of the current services for which it is responsible, to agree objectives and implement plans for future services, and to monitor the results of such provision. The examples given above are a few simple and necessarily fragmented illustrations of the relevance of linked records to management. They have been concerned only with linkage within hospital in-patient systems or between hospital records and mortality records. Brief mention must also be made of the possibilities of linkage between other sets of data—for example, linkage between community health and hospital records to trace the care received by people in different settings, or between records of clinical activity and finance to identify the costs of care—and their translation into health policy.

This chapter has considered the management implications of record linkage as distinct from its research applications, described in earlier chapters. However, management and research cannot be viewed as entirely separate activities. Management is dependent on research to help fulfil its roles. Epidemiological, clinical, and health care research are all essential to help management formulate plans for services to improve the health of the community it serves. An important role of management must be to foster research and innovation to achieve progress in medicine. Management of health care without research and innovation would be merely concerned with perpetuating existing patterns and levels of activity.

References

ACHESON, E. D. (1968). The incidence and prognosis of fracture of the femoral neck. *Report No. R17*. Unit of Clinical Epidemiology, University of Oxford.

ASHLEY, J. S. A. (1972). Present state of statistics from hospital in-patient data and their uses. *Brit. J. prev. soc. Med.* **26**, 135.

—— , HOWLETT, A., AND MORRIS, J. N. (1971). Case-fatality of hyperplasia of the prostate in two teaching and three regional-board hospitals. *Lancet* **2**, 1308.

BALDWIN, J. A., GOLDING, J., AND SIMMONS, H. M. (1980). *Linked record statistical tables, series IV: outcome* Vol. 1. Unit of Clinical Epidemiology, University of Oxford.

BENJAMIN, B. (1971). The use of statistics in the management of health and welfare services. In *Management and the Health Services* (eds. Gatherer, A. and Warner, M. D.) pp. 55–70. Pergamon Press, Oxford.

Department of Health and Social Security (1976). *Sharing resources for health in England: Report of the resource allocation working party*. HMSO, London.

—— (1977). *Hospital In-patient Enquiry* **iv**. HMSO, London.

—— (1981). *Orthopaedic services: waiting time for out-patient appointments and in-patient treatment*. Report of a Working Party to the Secretary of State for Social Services. HMSO, London.

GOLDACRE, M. J. (1976). Acute bacterial meningitis: where do children die? *Int. J. Epid.* **5**, 343.

—— (1981). Hospital in-patient statistics: some aspects of interpretation. *Com. Med.* **3**, 60.

LIPWORTH, L., LEE, J. A. H., AND MORRIS, J. N. (1963). Case-fatality in teaching and non-teaching hospitals. *Medical Care* (Philadelphia), **1**, 71.

MACKAY, D. AND CHARLES, J. (1951). *Hospital Morbidity Statistics*. General Register Office, Studies on Medical and Population Subjects No. 4, pp. 5–16. HMSO, London.

MORRIS, J. N. (1969). Tomorrow's community physician. *Lancet* **2**, 811.

Nuffield Provincial Hospitals Trust (1977). *Framework and design for planning: uses of information in the NHS* (ed. G. McLachlan). Problems and Progress in Medical Care. 10th series. Oxford University Press.

Office of Population Censuses and Surveys (1977). *Mental health enquiry*. HMSO, London.

19

Ethical implications: the protection of confidentiality

C. DU V. FLOREY

Introduction

The medical profession, followed at some distance by the general public, are showing increasing resistance to the linkage of data from various medical records using centralized government computers. In this chapter we look at the background to current concern, examine the principles of privacy, and see how these have affected the progress of linkage in the United Kingdom. We end with a commentary on ethical guidelines and how they may be met.

The issue of privacy extends well beyond the confines of medicine. Concern within Parliament was shown as long ago as 1961 (Lord Mancroft's Bill) and several attempts were made during the 1960s to introduce Private Members' Bills on the subject. During the debate on Mr Brian Walden's Bill in early 1970, the Home Secretary announced the appointment of a Committee to consider the problems of privacy in the private sector (the use of personal data by government was not within the remit). Two years later the Committee, under the chairmanship of the Rt. Hon. Kenneth Younger, submitted its report (Report of the Committee on Privacy 1972). It was extensive, clearly written, and contained comments on the lack of evidence of violation of individuals' privacy.

Public attitudes to privacy

The Committee considered that privacy included not only the protection of information but also other aspects, such as the powers of entry and search into the home and into personal and business affairs, and by media publicity. To test for attitudes among the general public regarding the notion of privacy, the Committee commissioned a survey through the Office of Population Censuses and Surveys. Although attitudes may have changed since then, the diversity of people's notions of privacy probably still applies. In the sample of 1596 responses nearly half the respondents saw the meaning of privacy as 'non-interference' or 'minding one's own business', and one-third felt it referred to home life and family or 'keeping one's own affairs to oneself'. Among 10 general issues (of which keeping down prices was the most important), protecting people's privacy ranked only seventh. Nevertheless, among seven social issues it ranked first, above improving race relations, protecting freedom of speech, or of the press, or equal rights for women.

A series of questions was asked about the publication of various personal details. One question asked whether the respondent would object to having his or her medical history made available to anyone who wanted to know. Surprisingly, since the profession gives such high priority to the confidentiality of the medical record, 49 per cent said they would not object (were they those who had no history?). Nevertheless, more than half the people did not like the idea of putting their medical history on public view, and 87 per cent of the respondents said that a central computer containing personal details should be prohibited by law. These findings and the Committee's statement that 'many of the anxieties which had led to demand for the creation of a legal right of privacy concerned the activities of government departments and public agencies', suggest that after more than a decade of public discussion of the privacy issue, the opposition to large governmental medical computer systems is likely to be substantial.

Confidentiality and new technology

The threat of computers to privacy is probably clearer in the public's mind now more than ever before as the capabilities of these machines are increasingly demonstrated in the office and at home. Three recent developments may further affect the privacy of individuals.

Firstly, full text retrieval is a system for searching stored text for specified items. For example, it is possible to enter newspaper articles over a period of time and then to search them for particular words and names. Thus, information may be linked rapidly in a way that would be extremely difficult using more conventional methods, or combined in novel ways prompted by the computer. The power of the system lies in its speed of retrieval from an unstructured data-set.

Secondly, word processors have become much more widespread in the last few years as their cost has fallen and the software improved. These machines are in effect computerized typewriters which record the words and spacing on a magnetic medium. Letters, prescriptions, hospital discharge summaries, and other documents can be stored for recall whenever required. Some systems have software for full text retrieval, so the scanning of stored summaries for data concerning patients could be easily achieved even by people who have little training or understanding of computers. If a comprehensive file of all typed work is kept in processor-readable form, then the potential for misuse may be very great indeed.

Thirdly, and allied to word processing, is image processing, in which digital representation of documents can be stored in a computer for reproduction or transmission to other sites. Although this has not yet become commonplace in hospital management, it must be expected to follow the rapid progress of word processing into the medical field.

The fear of a centralized computer and the reluctance of at least half of the respondents in the Younger Committee's survey to have their medical records divulged to others, must now be seen in a somewhat different light. The potential for interference with the confidentiality of the doctor–patient interchange, either by the casual browser or by those determined to obtain information about specific individuals, increases as each advance in computer technology enables more data to be stored at less cost. These developments raise two important ethical problems. The first is how the confidentiality, assumed by most patients, can be preserved in spite of the scanning of records by automatic data processing equipment, and the second is how confidentiality can be preserved in government data banks.

The principles of privacy

One may distinguish between privacy from intrusion and privacy of information about oneself. Privacy of information forms part of the wider concept of data protection and it is the latter which carries ethical implications for record linkage systems (and any other system in which personal data are gathered together). The principles of privacy of information set out the rights of the individual to control the use and dissemination of his personal data and provide boundaries on the behaviour of the users of the data.

Principles of data protection

Principles for data protection have been published by several groups and organizations, including the Younger Committee, the Organization for Economic Cooperation and Development (OECD), and the Council of Europe. Although there are differences in emphasis and wording there is also much in common. The Council of Europe's principles may be used as a starting-point (Council of Europe 1975). They are separately described for the private and public sectors but the main difference between the two is that control of the public sector should be by law, whereas control of the private sector is not specified, in recognition of the far greater power of government to invade privacy. The principles can be put under five headings:

(1) *Access to information about data banks*

In general, the public should have access to information describing existing data banks. This principle is concerned more with computer than manual data banks because of the difficulty of defining a manual data bank (a list of only 30 names under some disparaging title might be exceedingly damaging and could be construed as a data bank) and of the awesome task of recording all such collections. However, the principle is clearly appropriate for record linkage systems. Provisions might include a standard printed explanation which could be shown to every patient attending for medical care. For interested patients there should be a public register of health service computer systems which would describe the purposes of each system, the detailed code of practice by which it is governed, the types of data used, and a class-list of third party users allowed access to the data.

(2)   *Nature of the data*

The principles concerning the data themselves stipulate that the data must be accurate, up to date, stored for a defined time, and then erased. Furthermore, access to the data should be by the original user only. If taken literally, these principles present difficulties in the medical context, so flexibility in their application is necessary if they are to remain universal. For example, revisions of diagnosis or laboratory results should be recorded to keep the data as accurate as possible and up to date, whereas the need for an up-to-date telephone number or address code might be seen as less pressing. The length of storage may vary for the same data according to whether they are used for direct care of the patient, management of the health service, or for research. In the first and last cases it may be for an indefinite period, whereas management data may need to be kept only a short time. This problem is discussed later in the chapter. Flexibility is also required in the definition of user, since there may be multiple users within a hospital or general practice.

(3)   *Communication*

There are two principles of communication. Firstly, data files should not be passed to a third party without authorization. In those countries where legislation for privacy has included the establishment of a Data Protection Authority, independent of government, the Authority could give the required permission. In countries like the United Kingdom, where there are few legal safeguards for data privacy, the source of authorization is by no means clear. Were the institutions of the medical profession to have this authority, they could not be seen as disinterested in the outcome. Were authority to be vested in the government (the Department of Health, for example), the fear of government manipulation of data banks would not be allayed. At the time of writing, in the United Kingdom, there is no completely satisfactory solution to this problem.

The second principle of communication states that the data subject must have access to data about himself for the purpose of verifying their accuracy. Moreover, he should have the right to have errors corrected and to discover the way in which the data have been used. The most controversial part of the principle is where it implies the right of the patient to see his own notes. This is seen as a threat by some doctors as it would allow the patient access not only to what he already knows, but also to provisional diagnoses which may alarm him, and possibly to unsubstantiated conjectures about his motivation which, though they may help the doctor during the course of the patient's illness, could be damaging to the doctor–patient relationship if seen by the latter. However, this fear is mainly concerned with written notes and in most contemporary computer systems information will be limited by the space available for storage and the cost of abstracting, coding, and editing data from original records. The data are likely to be 'harder' and to exclude conjecture other than diagnosis. The likelihood that patients will demand to see their computer-held data, if given the right, will be very small judging from experience in other countries.

The issues raised by patient access pale into insignificance when placed beside the problem of identification of the applicant in countries like the United Kingdom where there is no mandatory identification document with a photo of the owner. The possibility of giving data to the wrong person, who may have his own nefarious use for the information, is far greater than is the chance of putting the doctor–patient relationship under stress. This is a problem universal to those in charge of data banks and one which has no immediate solution.

(4) *Security*

The principle of security is self-evident for any type of data bank, manual or computerized. It states that data banks should be equipped with security systems to prevent unauthorized people from gaining access to the information. This is more easily done for computerized records since the computer itself may allow or deny access to users according to their credentials. Manual systems would need to be securely locked to prevent browsing, although an open filing system can be designed to prevent easy identification of a specific person's documents.

(5) *Legitimacy*

Finally, the principle of legitimacy states that data must not be collected by fraudulent means or used for illegal purposes. Medical ethics would insist that this principle be upheld in the collection and use of medical data. However, in addition, it implies that the principle involving the right to know how the data are to be used has been accepted and acted upon. Patients are usually content for their manual records to be stored in large hospital filing systems, believing, probably correctly, that purposeful access to the notes by unauthorized people is unlikely or too difficult to be worthwhile. But they will be aware of the ability of a computer to sift data swiftly and may be more anxious about the use to which their data will be put when stored in machine-readable form. To meet the principle of legitimacy the patient must know what will happen to the information he supplies to the doctor concerning his history and physical examination. Within a hospital, the system for informing the patient comes under a single administration; for record linkage the system is vastly more complex because of the many sources of data, but the principle gains in importance because of the greater potential for damage from misuse of linked records. The principle of legitimacy is not discussed further here as in itself it poses no theoretical or practical problems, but its co-principle will be considered again in formulating recommendations.

The National Health Service, research, and the privacy principles

The Younger Committee considered the problem of privacy in medicine but, because of its limited remit, it could not include the National Health Service (NHS). It was believed that, by and large, privacy was adequately protected as there were few complaints about doctors in the evidence given to the Committee. Comments on clinical and epidemiological research were equally encouraging. The protection of data privacy in government was not opened for discussion until the White Paper, Computers and Privacy (1975), was published in December 1975 in which an outline government policy for data protection was described. Following this, the Home Office established the Data Protection Committee under the chairmanship of Sir Norman Lindop, to recommend a feasible scheme for legislation for the protection of data processed by automatic means. The Committee sat for two years, during which time much written and oral evidence was reviewed. Medical evidence was disproportionately greater than that of other interested parties and it was treated under the heading of 'Areas of special concern' in the Report (Report of the Committee on Data Protection 1978).

Recommendations of the Data Protection Committee

The Committee listed the principles to be declared in the Statute. These excluded the principle of subject access because the evidence, mainly from medical practitioners and research workers, made it clear that it would be very objectionable under some circumstances and unnecessarily expensive in others. For example, if there were a right of subject access to research records, it would be necessary to include in each project a method for retrieving an individual's data, even if no request was ever made. The potential cost of this service would have to be written into each grant application, thus inflating project budgets and reducing the funds available for other investigations. The alternative suggestion was that a Data Protection Authority (DPA) should ensure that the appropriate data were held and that they were used legitimately. In some cases this would mean direct subject access, in others possibly inspection by the DPA in the event of a complaint.

Despite these limitations on the application of the principle of subject access, there was adverse reaction from some medical quarters, particularly the British Medical Association (BMA) and some of its publications. The Central Ethical Committee (CEC) of the BMA prepared three principles for confidentiality (British Medical Journal 1978), which were primarily concerned with the protection of identifiable medical information from unauthorized third parties, including research workers and those involved in the preparation of statistics. The first two principles give the patient control over his records to the extent that his consent is required for access by people other than those concerned with his immediate care. Nevertheless, the clinician may also grant access, on his own authority, provided it is in the patient's interest. The patient has absolutely no right of access to the notes himself. The third principle states:

'An individual is not to be identifiable from data supplied for statistical or research purposes except when follow-up of the individual patient is a necessary part of the research and either the patient has given informed prior consent, or consent has been obtained from the chairman of an appropriate ethical committee.'

Although in many ways the principles are laudable, they present particular difficulties for record linkage systems and epidemiological research.

One of the advantages enjoyed by British epidemiologists but unavailable to colleagues in many other countries, has been the linking for research purposes of birth and death records with other data obtained elsewhere. The third BMA principle rightly stresses the appropriateness of disidentification of personal data before the data are handed to bona fide third parties who have no need for the identification. But for projects where linking is essential, there is no body independent of the medical profession which can judge the merits of the proposed safeguards for confidentiality. The only bodies that might give guidance at present are District ethical committees, which are unlikely to have the experience to make a sound judgment, the CEC of the BMA, which is strongly influenced by BMA policy, or the General Medical Council, which, to date, has opted to stand on the sidelines concerning the confidentiality of medical records held in computers. None of these could be seen by the general public to be totally disinterested.

The Child-Health Computing Committee code of practice

Since there was no recognized authority, an impasse developed between the Department of Health and Social Services (DHSS) child-health computing system and the CEC. In order to rationalize NHS computing, the DHSS aimed to have a standardized computing system throughout England and Wales rather than leave each Area or Regional Health Authority to work out its own separate (and probably incompatible) system. The objective was to record information, some of it sensitive, about the mother and child from hospital, community services, and general practice in order to monitor immunization and developmental screening programmes and so on, and to prompt action and ensure appropriate care. The Child-Health Computing Committee, representing the NHS and professions, was set up in 1977 to advise centres responsible for the development of the computer-based child-health system and to co-ordinate and monitor the developments as necessary. The Committee accepted the three principles of the CEC and gave its Confidentiality Safeguards Working Party the task of drawing up a code of practice based on them.

The code of practice for the system separated security from confidentiality. Security refers to the physical integrity of the installations where the records are kept and processed, and the methods used to prevent unauthorized people from gaining access to either manual or computer-held records by whatever means. Confidentiality refers to the limitations the data subject wishes to put on right of access to information in his records and on the use to which that information may be put.

The Working Party's recommendations for security for the manual records were far more stringent than are commonly used in hospitals; for example, 'records should not be left unsupervised in places accessible to unauthorized persons'. Although recommendations for even greater security might have been suggested, they would have set a standard that few systems could meet at reasonable cost and would imply a radical alteration to current record-keeping in the NHS.

Confidentiality poses an altogether different problem within medicine since this is most likely to be broken, albeit in many cases inadvertently or innocently, by people working within the system. The sanction of instant dismissal may be placed on employees, but there is a significant number of voluntary workers involved in services to patients for whom no obvious sanction exists. For this group the Working Party's recommendation was only to exercise care in the selection and supervision of volunteers.

Levels of access

Four levels of access to data were described by the Working Party. Access for patient care would be restricted to the responsible clinician and those authorized by him. Access for legal purposes would have to comply with the law. Access for research purposes would be granted by the Area Medical Officer, as custodian of the data, for non-identifiable information or identifiable information which had neither medical nor social content (little would come under this heading since even address and telephone number carry social information). Access to identifiable information with social or medical information for follow-up studies would require permission from the custodian, and either the patient's consent or the consent of the local ethics committee. A signed declaration concerning the use of the data would be required from the research worker.

A separate paragraph in the recommendations was devoted to access for management purposes in which the rules were similar to those for research use, except that identifiable information with medical and social content was not seen to be relevant.

Finally, the Working Party recommended that a register of accessors should be kept by the custodian, to be made available to authors of the information, researchers, and other bona fide enquirers. Access to the register by patients was not mentioned.

This code of practice raises important issues because it leaves loopholes and its credibility depends on the users policing themselves. Some of the issues may have arisen because privacy in medicine has been dominated by medical thinking rather than a holistic approach in which other threats to privacy are considered. Although there are many special problems for medical records, the principles for data protection should be the same as in other fields; it is only their application which might vary to take into account particular circumstances.

The implications of government-controlled data banks

The Data Protection Committee saw the greatest potential danger to data privacy in the build-up and expansion of government-controlled data banks. At present, the creation of a massive central file is prevented by agreement between departments not to exchange data, by the difficulties of linking and merging data from different files in the absence of a universal personal identification number, and from the incompatibility of the data formats. However, these factors only make the merging of the data more difficult. They would not prevent it if a government were sufficiently determined to set up a central data bank. It is this threat, clearly seen by both the Younger and Data Protection Committees, which has not been considered in the medical debate. It makes a voluntary solution to the problem of confidentiality more difficult and increases the need for an authority, independent of users (including government) and with the means to enforce its decisions.

The larger problem of government-controlled data banks requires a somewhat different approach to that provided by the CEC in its three principles. Firstly, the definition of identifiable data needs to be much clearer. Disidentified data have so far implied data without name or address. But it is possible to identify people from other characteristics, such as height, ethnic group, social class, and so on. Thus the removal of name and address, which might be sufficient to prevent revelations to the browser, would be insufficient to prevent identification by skilled investigators. The Data Protection Committee opted for a wider definition, namely that personal information is any information which relates to any data subject who is or can be identified—taking guidance from definitions used in the privacy legislation of other countries. Personal data therefore cannot be disidentified, they can only be made difficult to identify. Under this definition, the privacy principles must be applied to all personal data whether or not overt identifiers are present.

Secondly, the larger problem also leads to a classification of data according to whether it is to be used for administrative or research purposes. Medical records, records of patients' admissions and discharges, and other records of health service activity would be classified as administrative, whereas files for research into data abstracted from these records and of data obtained in surveys would be under the 'research' heading. A new principle can be applied which states that personal data collected for administrative purposes may be used for research, but that personal data collected for research may never be used for administrative purposes. In this way, administrative data passed on for research, or data collected in studies of sensitive issues, can be protected from potential administrative misuses that might be to the detriment of the data subject. Because of the one-way traffic from administrative to research, any data not collected purely for research purposes must be dubbed administrative. Nevertheless, research would include the analysis of data to provide statistical summaries (from which identification of individuals was made impossible) even though the summaries might be used in the making of administrative decisions.

The third point which has been bypassed in the medical debate is the application of the principle that data should be stored for a defined time and then destroyed. Because the debate has focused on doctors' records, it has been assumed these records would be kept indefinitely, even after the death of a patient. But the larger problem requires consideration of this principle because systems such as that for child health, have the potential for gathering administrative data for all people born after its inception, over an indefinite period. The system has the capacity for becoming, in time, a vast medical data bank for the total population, which can be centralized with little effort because of the compatibility of the sub-systems throughout the country. Prevention of this possibility is precisely what the protection of data privacy in the public sector is about.

An outline code of practice for record linkage systems

The principles of data privacy are equally applicable to manual and computer-held records. However, privacy is threatened in different ways in the two methods of storage and therefore, a code of practice must take into account whether one or both methods are to be used in a particular class of application. Record linkage systems will inevitably involve the use of computers and this is acknowledged in the following guidelines.

The guidelines below are concerned with all personal data, whether overtly or covertly identifiable, except where otherwise indicated. They are primarily dependent on the internationally accepted principles of privacy but they also include safeguards to meet the principles given by the CEC of the BMA and particular points raised in the Medical Research Council's code of practice (Medical Research Council 1973). They assume that a named individual, the custodian, is responsible for the security and confidentiality of the data. In countries where there are no legal sanctions for failure to respect confidentiality, the custodian should be medically qualified so that the sanction of deregistration may be used.

Security

Manual records should be designed so that overt or plain text identifiers (name and address in most cases) can be separated from the rest of the data.

A numerical system can be used to link the two documents. The data on the identification documents should be transferred to computer-readable form and the document destroyed. Where this is not possible, the document must be kept in a high security area. When data are sent for transfer, the dis-identified document should be processed first to prevent association between name and the sensitive data.

There are a variety of ways for maintaining security of computer-held data which will not be discussed here. The highest security compatible with the budget available should be sought but, at the least, direct access to the data should be allowed only to named individuals.

Public register

There should be a public register of all computer systems in which personal data are handled by the health service. This should record the existence of each system, the code of practice used, the types of data held, the uses to which the data are put, and the classes of third party accessors. Among others, this would be open to patients who wanted to know in greater detail about the handling of their personal data. If there is no public register, a description of each record linkage system should be readily available to those who want it.

The data

Patients should be aware that their data are likely to be processed in a computer for administrative and research purposes and they should be told, in general terms, what safeguards there are for the data. More detailed description of the safeguards should be available from the register of applications. The patient should have the option not to participate in computer schemes in which linkage is by name, date of birth, or other overtly distinguishing data and which are not directly connected with his care. It would be expected that few patients would avail themselves of this right since it could restrict research into their illnesses, and there should be sufficient safeguards in the system to allay fears of breaking confidentiality.

Access to and communication of the data

The data in any given system are the responsibility of a custodian who should be medically qualified, unless privacy legislation provides the necessary sanctions for a custodian without medical qualifications. He should have an advisory committee, which may be the local ethical committee. He will deal with two classes of data, administrative and research, of which adminitrative data can be divided into those used for the direct care of the patient and those used for management of the health services.

Where access to the data is through a computer terminal or from printed output, overtly identifiable data for the continuing care of the patient should only be available to the doctor in charge of the patient. Direct access should not be delegated by this doctor to any non-medical personnel. Once he has acquired the data, he becomes responsible for protecting confidentiality and limiting its dissemination to those involved in the care of the patient. Release of overtly identifiable data by the custodian to a third party requires consent from the doctor in charge.

Data for management purposes should have overt identifiers removed and should only be accessible to employees within the system, as designated by the custodian. Different levels of access may be defined according to which data are required by different employees. Data requested by government departments, except where the law requires it, should only be provided when the custodian, in conjunction with his advisory committee, has satisfied himself that the receiving department has adequate confidentiality safeguards of its own.

Data for research workers outside the system should be supplied only to those known to be bona fide and associated with a reputable (in terms of protecting confidentiality) research institution. Overtly identifiable data which are only to be used for linking with other research data already obtained, may be provided to medically qualified researchers, provided the custodian and his advisory committee are satisfied that the use is proper and the safeguards adequate. If the patients are to be approached by the researchers, this must be done through each patient's doctor after the doctor has given the custodian permission to make available the overtly identifiable data.

No access to the records may be granted for administrative purposes other than those already declared. For example, the records should not be voluntarily opened to the police even if it were to assist in their investigations.

At present, in the United Kingdom, patients should not have access to their own computer file or manual records unless the doctor in charge has given permission. However, in the event that a patient has a genuine belief that incorrect data or unsubstantiated judgmental data are held about him, which may be or have been used to his disadvantage, then access should be granted to an independent person, nominated by the patient and acceptable to the custodian, in order to check the correctness of the records and to verify that corrections have been made. This type of access is only likely to be used if the data have been released for improper administrative purposes (see above).

Entries to a record linkage system should be dated. Data should be stored indefinitely for research purposes. The storage should be in computer-readable form and overt identifiers should be physically separated from the rest of the data. The data should be available for management purposes for a defined period after collection. The data may be kept for a different fixed period under the administrative heading of direct care. The length of the periods will depend on the uses of the data declared in the public register.

Administration of a code of practice

This outline code of practice for record linkage systems within the health service may help as a guide to ethical operations but cannot deal with the fundamental problem of control. In the interests of the data subject, codes of practice should be approved by bodies such as the GMC on which lay members as well as medical members sit. With experience, they might act in the way that the Data Protection Committee described the working of a data protection authority, helping to write the codes of practice with the users and approving changes to meet changing circumstances.

Alternatively, restructured local ethical committees might take on this function. These committees should consist of people with experience in the field of data protection, including lawyers, employers, trade unionists, statisticians, experts in automatic data processing, and public administrators. They should also have members of the public who are known to be concerned with the rights of patients, such as those serving on Community Health Councils in the United Kingdom. Clinicians and research workers should also participate but their role might be seen less as one of protecting the data subject than of bridging the gap between the applicant or user and the other members of the committee.

If problems arise in new applications, the committees should work with the users to write a mutually acceptable code of practice. If, after this process, an application were still unacceptable to a committee, the user should have an avenue of appeal which might be to a national medical ethical body, such as the GMC. The committees should also investigate legitimate complaints from the public. These may lead to modification of the code of practice to prevent further problems or, if the breach of confidentiality were sufficiently serious, the incident might be taken to the GMC for disciplinary action.

Legislation

But however well a profession may police itself, it may be suspect in the eyes of the public and will be powerless to enforce its rules on an antagonistic government. The practical problems of ensuring confidentiality of personal medical data will not be resolved to the satisfaction of the majority of concerned people unless there are clear legal obligations, for both the private and public sectors, and there exists an overseeing body whose prime objective is to protect the individual.

In 1982, the Conservative government published another White Paper on privacy legislation (Data Protection 1982), three years after the Lindop Report. It recommended that an independent Registrar be appointed by the Crown. He would be responsible for the compilation of a public register of all systems processing identifiable personal data by automatic means. There would be no code of practice enforceable by law, contrary to the recommendations of the Lindop Committee, but users would be expected to abide by the spirit of eight general principles similar to those described in this chapter. There would be criminal and civil sanctions for violations of the principles. In some areas, the principles would need to be supplemented by regulations. Some of these categories of data (in particular, medical records) were expected to be covered by regulations which might, for example, place special restrictions on the collection, processing and holding, or disclosure of information from such records. Who was to make the regulations and to what extent they would differ from the concept of a code of practice, was not made clear. Additionally, the control of data systems in the public sector was dealt with less explicitly than private sector controls, leaving loopholes for subtle misuse of personal data.

In 1983, the Data Protection Bill was published, based on the White Paper. The recommendations of the Lindop Committee—to set up a Data Protection Authority and to develop codes of practice—had not been accepted. The Bill left a number of questions unanswered, including how special regulations might be created, what records came under its influence, how a data subject might gain access to his own records, and whether there were adequate safeguards for restricting transfer of data within the public sector. Of particular concern to the medical profession was the potential for access to medical records by people to whom they are not usually available. The provisions of the concurrently considered Police and Criminal Evidence Bill increased the profession's unrest. The general election of June 1983 halted both bills, but they were introduced again after the re-election of the Conservative Party. By July 1984, both the Data Protection Bill and the Police and Criminal Evidence Bill had been enacted, and the framework within which personal information in the National Health Service is to be handled was crystallized.

References

British Medical Journal (1978). Child-health computing system. **1** , 1432–3.
Computers and Privacy (1975). Cmnd 6353. HMSO, London.
Council of Europe Resolutions (73)28 and (74)29 (1975). (The resolutions are set out in Tables 8 and 9 of the White Paper Supplement, Cmnd 6354. HMSO, London.)
Data Protection (1982). The Government's proposals for legislation. Cmnd 8539. HMSO, London.
Medical Research Council (1973). Responsibility in the use of medical information for research. *British Medical Journal* **1** , 213–16.
Report of the Committee on Data Protection (1978). (Chairman: Sir Norman Lindop.) Cmnd 7341. HMSO, London.
Report of the Committee on Privacy (1972). (Chairman: Kenneth Younger.) Cmnd 5012. HMSO, London.

20

Future directions for medical record linkage
THE LATE J. A. BALDWIN

This textbook has necessarily been concerned almost entirely with the achievements of medical record linkage, but it is salutory to end it by reviewing briefly what has not been accomplished and indicating some possible future developments which now seem both desirable and practicable.

In his classic *Medical record linkage*, Acheson (1967) devoted some 20 pages to describing ways in which linked records might be applied to a wide range of medical issues, and in the Introduction to this book he has highlighted the most important applications and developments over the last 20 years. The general impression is that work has been done in virtually all the fields originally envisaged. There are abundant examples of the use of linked hospital in-patient and vital records to prepare statistics on persons rather than episodes, yielding measures of incidence, prevalence, readmission, re-operation, and survival rates. There have been useful contributions to the major fields of mental health and reproduction, which Acheson singled out for special mention. The study of disease associations and morbidity consequent on previous care has been advanced. There have been developments, albeit limited, towards means of follow-up at the national level, and experiments in monitoring industrial hazards and for adverse effects of drugs. Linked records systems, like those in the Oxford Region and in Scotland, are used frequently for case ascertainment, for personal follow-up, for the identification of groups at special risk, and for such mundane but important tasks as death clearance in medical records systems.

Yet even the convinced reader will experience a sense of unease. He will note that no single system has developed all the areas of application, only one major new system (the OPCS Longitudinal Study) has been created in the United Kingdom, and none of the more ambitious schemes for national systems, such as for follow-up, drug monitoring, or detection of industrial hazards, has yet evolved in England and Wales. There have been many useful and even important contributions to medical knowledge and probably some effects on medical care, but no dramatic discoveries that have caught the imagination of the medical profession, much less of the general public or of the administrative powers. Except in scientific applications, progress seems to have been slow and patchy.

The obstacles to progress

The reasons for this slow progress are not hard to find. Throughout this formative period, despite evident technical solutions to the problem of linking records reliably, the technological resources were not well suited to handling the very large data files which were accumulating. Computing technology was changing so rapidly that systems development was hardly completed before the hardware was replaced by more powerful but incompatible new machines, and their cost was so high that record linkage systems could not justify sole ownership. At the national level particularly, economic, political, and ethical opinion became less favourable so that progress had to be made with the utmost caution and circumspection if it were not to be stopped altogether. Misconceptions were common about the nature and purposes of medical record linkage. It was alternatively academic and irrelevant or mysterious and dangerous. Health service computing staff remained largely unaware of the potential for applications in their hospital systems.

A curious view grew up among administrators in the early 1970s that linking records was very expensive and could not possibly be justified for this reason alone. This seems to have arisen from a confusion between the cost of collecting information for such systems as hospital activity analysis and the added cost of linking it. In Chapter 2 it was emphasized that the marginal cost of collecting the necessary identifying data and of linking records collected for use in unlinked form is trivial in relation to the increased value of the information that can be extracted. Yet it remains true that the linking process and the use of linked information reveals depths of error and inconsistency which escape even the most advanced editing techniques available for episodic data. There is a cost in bringing health service data up to a standard to make them worth linking which, it may be argued, ought to be met even for applications in unlinked form. A further cost which is rarely recognized and is therefore also confused with the cost of linking, is that incurred in developing techniques, notably software, for analysing and presenting linked record statistics in a usuable fashion. Some of this development has been completed and would-be users of linked data may benefit from the methodologies described in Chapter 3.

In recent years, resource constraints have further hampered progress in some systems. An unfortunate effect is to deter health service administrators and managers from using linked data. There is a strong belief, often unjustified by the evidence, that unless statistical intelligence is absolutely up to date, it has no more value than yesterday's news. Medical record linkage systems cannot remain indefinitely and exclusively in the academic and research fields, insulated from the realities of the health services. In particular, they cannot escape the burden of timeliness if they are ever to appeal to health service managers.

Apart from these largely circumstantial reasons for the limited impact of medical record linkage, it is in the nature of the subject that its full potential is slow to be realized. There are three interrelated impediments to rapid exploitation which affect linked medical record systems in varying degrees. One, which is common to all, is the necessity for sufficient time to elapse for outcomes to be detected. The latent periods of many pathological processes may span several decades. Acheson (1967) cites 30 years before the onset of clinically detectable disease following exposure to toxic anti-oxidizing agents used in rubber manufacture and even longer periods are suggested for certain cancers. Since the mean period at risk in a constantly accumulated linked file is half the lifetime of the file, numbers of cases sufficient to obtain reliable results are unlikely to be available in files of manageable size until some time beyond the latent period of the condition under study. The relation between diseases occurring early in life and those maximally incident in late life, such as schizophrenia and cancers, present analogous difficulties.

The limitation of elapsed time is associated in smaller systems with the impediment of scale. Many of the phenomena which it is desirable to observe in linked data are relatively infrequent—the expected congruence of two quite common events in the life of an individual is surprisingly small—so that what is observable is related to the size of the population covered by the linked record system. Very large systems, like the Scottish example with a population at risk of about five million, are much better placed in this respect than miniature systems with a catchment of, say, 500 000 population. The Oxford Study has a maximum follow-up only in its starting population of about 340 000, and even with its coverage now approaching two million after 20 years, the mean follow-up is only about 7.5 years over the linked data-file as a whole.

The third impediment is the data content. Items of information not included in the early records cannot be used for sample selection for the longest follow-up periods without undertaking difficult and costly special collection. Thus the starting data-set must be designed with the greatest care and prescience. Yet the temptation to include too many not immediately usable items must be resisted, partly on grounds of cost, but mainly because the reliability and completeness of collection is likely to be adversely affected by disuse coupled with suspicions of 'irrelevance'. Furthermore, the larger the system the more difficult it becomes to maintain adequate standards of collection, and paradoxically, the gain in numbers from large scale may be partially offset by the paucity or unreliability of data. In the Introduction, Acheson expresses regret that information about drug treatment was not included in the starting data-set for the Oxford Study so that investigation of the efficacy of specific therapies and monitoring for adverse effects could not be undertaken. To this may be added the limitation imposed by the absence of any record of previous history. In this sense the Oxford data-file has no beginning since, in the presence of the high frequency and ubiquitousness of readmission, it is not practicable to derive certain incidence rates with con-fidence; in studies of chronic conditions it is usually necessary to ascertain by scrutiny of clinical records whether the first encounter mentioned in the linked data is really the first occasion on which the diagnosis was made. Although this limitation is, in principle, self-correcting with the increased lifespan of the file, it will be many years before its effects cease to be of importance.

Future developments

The obstacles to more rapid and extensive exploitation of linked medical records suggest directions for future development. The procedures for linking records, though undoubtedly capable of much further improvement, are no longer a barrier to widespread use. They are cheap, reliable, and efficient and can be implemented on a considerable range of modern computers. The primary task should be to ensure that existing schemes are kept going long enough and on a sufficient scale to yield the dividends now due on the investment in them. No one system is likely to be able to attract the resources, skills, and interests to embrace all the possible uses of linked records, and new schemes designed to explore underdeveloped applications ought to be encouraged. New applications can be foreseen in the scientific, clinical, and health service management fields.

Scientific applications

There are already strong indications of the importance of the OPCS Longitudinal Study. The work of Fox and his colleagues on mortality (Chapter 4) has more than justified this system and has emphasized the continuing epidemiological value of mortality studies which the Office of Population Censuses and Surveys is uniquely fitted to undertake. Longer-lived systems, such as those in Scotland and in Oxford, are ripe for extensive investigations of outcome of treated illness in far greater depth than has so far been attempted and these should go forward in parallel with more conventional case-control studies, trend studies, and so on. The work begun in Oxford on disease associations is still in its infancy and should yield more important and reliable results as the temporal and population scales increase.

One very large area which remains virtually unexplored by means of linked records is that concerned with the family and genetic factors in disease. The wider recognition of the importance of genetic and familial influences in recent years should stimulate interest in the use of linked data to build upon the classic work of Newcombe (Chapter 14). The practicality of family record linking in systems which do not have the advantages of the Canadian vital records has been limited, but with the increasing numbers of records of completed families accrued through obstetric data collection and more powerful identifying data-sets, there will be a progressive improvement in what can be done. Human geneticists may find linked record systems a more useful source for sampling and selection of pedigrees than has been available hitherto.

Contrary to what may be achieved in Canada and the United States, the prospects for effective systems for monitoring industrial and drug-treatment hazards in the United Kingdom seem distant. Demonstrations of feasibility such as that by Skegg and Doll (Chapter 10) have not been sufficient to divert the resources necessary for implementation—perhaps we must suffer the agony of another Thalidomide disaster to invoke action.

Clinical applications

Record linkage has not had a direct impact on the clinical field. The habit of thinking in terms of people rather than episodes of illness, of long-term outcome rather than the immediate effects of hospital care, has yet to become widespread. It is possible to envisage change if timely feedback of linked data can be developed for such purposes as service management, follow-up, and audit. As more robust and refined statistical estimates of outcome are extracted, quantitative prognosis should become a reality.

Health service management applications

Record linkage is a tool of immense versatility. The wide range of applications in medical research exemplified in Part II of this book is but one aspect of a multi-faceted jewel. In Chapter 18, Goldacre gives pertinent guidance on how linked data can enhance management intelligence in the health services, and similar kinds of material would integrate with clinical uses. Beyond these statistical applications, the technique of record linking has an important place in the day-to-day operation of the new District health services. Name matching and record linking technology can be used efficiently in District computing systems. Indeed, this will become necessary if District management teams are expected to make use of intelligence based on the person as the unit of record rather than the hospital in-patient admission, following recommendations of the national Körner Committee (Steering Group on Health Services Information 1981).

The existing hospital activity analysis system, whether or not it evolves into the more comprehensive patient information system advocated by this committee, may have to derive District episodes based on a District person number allocation system. At present, very few large hospital master indexes, whether or not automated, are likely to be as reliable as properly designed name-matching procedures in constructing person or 'unit' records (Baldwin and Gill 1982). Development into larger District master indexes, using existing index identifying data-sets and relatively unsystematic search and match procedures, would be likely to reduce reliability further. Satisfactory efficiency and reliability of operation will require incorporation of a full probability-match routine in the computing software, together with some enhancement of the identifying data-set. On-line operation in this way now appears to be practicable.

This application would open up considerable potential for use of linked data since all hospital registrations would be accessible for statistical purposes rather than only in-patients. Pathology, X-ray, pharmacy, and eventually financial data might be encompassed as hospital computing systems are developed. Some Districts are interested in links with primary care through Family Practitioner Committees thereby broadening the scope for collaboration with general practice use of morbidity data.

Developments of this kind would have only limited direct value for epidemiological work because of the small scale of the District. Linkage to higher levels of aggregation would still require special arrangements, such as those for the Oxford Study and in Scotland. Nevertheless, provided that adequate arrangements are made for data protection so that administrative users, as defined by Florey in Chapter 19, are allowed access only to statistical material, and transfers for medical research use continue to be subject to the present stringent safeguards, there would be scope for benefits to the public health as well as to the quality of the health services.

The importance of this wider application of record linking technology can scarcely be overstated. It would be a major new direction which would establish medical record linkage within the health care system, signifying the end of the experimental phase. Many of the practical constraints experienced hitherto would be overcome and its future would be assured. With this achievement medical record linkage might truly be said to have come of age.

References

ACHESON, E. D. (1967). *Medical record linkage*. Nuffield Provincial Hospitals Trust. Oxford University Press.

BALDWIN, J. A. AND GILL, L. E. (1982). The district number: a comparative test of some record matching methods. *Community Medicine* **4**, 265.

Steering Group on Health Services Information (1981). *Report of the Working Groups A*. HMSO, London.

INDEX